Data Representations, Transformations, and Statistics for Visual Reasoning

Synthesis Lectures on Visualization

Editor
David S. Ebert, *Purdue University*

Synthesis Lectures on Visualization will publish 50- to 100-page publications on topics pertaining to scientific visualization, information visualization, and visual analytics. The scope will largely follow the purview of premier information and computer science conferences and journals, such as IEEE Visualization, IEEE Information Visualization, IEEE VAST, ACM SIGGRAPH, IEEE Transactions on Visualization and Computer Graphics, and ACM Transactions on Graphics. Potential topics include, but are not limited to: scientific, information, and medical visualization; visual analytics, applications of visualization and analysis; mathematical foundations of visualization and analytics; interaction, cognition, and perception related to visualization and analytics; data integration, analysis, and visualization; new applications of visualization and analysis; knowledge discovery management and representation; systems, and evaluation; distributed and collaborative visualization and analysis.

Data Representations, Transformations, and Statistics for Visual Reasoning
Ross Maciejewski

ISBN: 978-3-031-01471-0   paperback
ISBN: 978-3-031-02599-0   ebook

DOI 10.1007/978-3-031-02599-0

A Publication in the Springer series
SYNTHESIS LECTURES ON VISUALIZATION

Lecture #2
Series Editor: David S. Ebert, *Purdue University*
Series ISSN
Synthesis Lectures on Visualization
ISSN pending.

Data Representations, Transformations,and Statistics for Visual Reasoning

Ross Maciejewski
Purdue University

SYNTHESIS LECTURES ON VISUALIZATION #2

ABSTRACT

Analytical reasoning techniques are methods by which users explore their data to obtain insight and knowledge that can directly support situational awareness and decision making. Recently, the analytical reasoning process has been augmented through the use of interactive visual representations and tools which utilize cognitive, design and perceptual principles. These tools are commonly referred to as visual analytics tools, and the underlying methods and principles have roots in a variety of disciplines. This chapter provides an introduction to young researchers as an overview of common visual representations and statistical analysis methods utilized in a variety of visual analytics systems. The application and design of visualization and analytical algorithms are subject to design decisions, parameter choices, and many conflicting requirements. As such, this chapter attempts to provide an initial set of guidelines for the creation of the visual representation, including pitfalls and areas where the graphics can be enhanced through interactive exploration. Basic analytical methods are explored as a means of enhancing the visual analysis process, moving from visual analysis to visual analytics.

KEYWORDS

visual analytics, histograms, scatterplots, parallel coordinate plots, multivariate visualization, power transformation, time series analysis, choropleth maps, clustering

Contents

Acknowledgments

I would like to thank David Ebert, Jason Dykes, Diansheng Guo, and William Ribarasky for their helpful discussions in preparing this manuscript.

Ross Maciejewski
May 2011

CHAPTER 1

Data Types

Visual analytics has been described by Thomas and Cook [128] as the science of analytical reasoning facilitated by interactive visual interfaces. These interfaces utilize combinations of statistical graphics, analysis, animation and interaction in order to enhance the cognitive process and provide users with an intuitive means to explore and understand their data. In this way, users can confirm expectations about their data, search for the unexpected, formulate hypotheses and utilize tools to help them express the story being told within their data. The visual analytics process should not purely be a visual exploration of the data where plots and graphics provide users with summaries of the data, nor should visual analytics be a purely analytical process where the data is sent through a multitude of machine learning algorithms and a set of anomalies returned. Instead, visual analytics should work to combine these processes together in a seamless manner to enable the discovery, generation and testing of hypotheses in an interactive framework. By combining analysis and data preconditioning early prior to visualization, visuals that guide users to obvious issues in the data can be created. Through the use of interactions, the obvious needles in the haystack can be discovered, analyzed and then pulled out of the haystack. From there, new searches can begin that tease out pieces of information that may be hidden. This process of teasing out information needs to loop back to the underlying analytics as a means of evaluation and hypothesis testing.

During the creation of visual analytics tools, we need to be cognizant of design parameters not only in the modeling and analysis of the data but also of design parameters for visual representations. Appropriate parameter choice can enable analysts in exploring and analyzing their data, while poor parameter choices can obfuscate and even mislead users. In fact, all visualization is subject to design decisions and many conflicting requirements. Visual analytics allows you to vary design parameters to suit a particular need, and do so rapidly and interactively as needs change. These tools have their own strengths and weaknesses, and as the amount of data being stored and process increases, the need for tools to facilitate the analytical process is ever increasing. In order to understand the visual analytics process, it is necessary to understand the common visual tools and analytical techniques that can be used as a basis from which to develop visual analytics systems.

1.1 DATA TYPES

Analysis first begins with a collection of data sources. These data sources can be multi-scale and multi-sourced in which an analyst will go through a process of cleaning, transforming and modeling data with the goal of extracting useful information. This information is then used to develop conclusions and support decision making.

In order for data to be ingested into computers, it needs to have a structured form suitable for computer-based transformations. These structured forms exist in the original data or are derivable from the original data. Structures retain the information and knowledge content and the related context within the original data. These structures are transformable into lower-dimensional representations for visualization and analysis. For example, in Table 1.1, a sample dataset is given consisting of the top twenty players (based on their batting average) in Major League Baseball's National League. Each column represents a data structure containing details about players and their output within Major League Baseball.

Table 1.1: Baseball stats of the top twenty players (based on batting average) in the Major League Baseball National League, 2009-2010 season.

Name	Team	At Bats	Runs	RBI	Batting Ave
C. Gonzalez	COL	587	111	117	0.336
J. Votto	CIN	547	106	113	0.324
O. Infante	ATL	471	65	47	0.321
T. Tulowitzk	COL	470	89	95	0.315
M. Holiday	STL	596	95	103	0.312
A. Pujols	STL	587	115	118	0.312
M. Prado	ATL	599	100	66	0.307
R. Zimmerman	WSH	525	85	85	0.307
R. Braun	MIL	619	101	103	0.304
S. Castro	CHC	463	53	41	0.300
H. Ramirez	FLA	543	92	76	0.300
P. Polanco	PHI	554	76	52	0.298
A. Gonzalez	SD	591	87	101	0.298
J. Werth	PHI	554	106	85	0.296
M. Byrd	CHC	580	84	66	0.293
A. Ethier	LAD	517	71	82	0.292
A. Pagan	NYM	579	80	69	0.290
A. Huff	SF	569	100	86	0.290
J. Keppinger	HOU	514	62	59	0.288
D. Uggla	FLA	589	100	105	0.287

In this case, the structure of the data is already provided in terms of the rows, columns and linkages to player names and teams. However, in the case of data such as text documents, video and pictures, the structure of the data is not always so readily available. In those cases, analysis is often done to determine ways to link these sources of unstructured data to known structured data. While there may be inherent structure that can be inferred from video, text, etc., the fact that the main content being conveyed does not have a defined structure is what classifies such data sources as undefined. In this chapter, the focus is solely on structured data types and analysis.

Once a data structure is defined, the goal of visual analytics is to choose an appropriate visual representation to translate data into a visible form. Visual representations make it easy for users to perceive salient aspects of their data quickly. These visual representations augment the cognitive reasoning process with perceptual reasoning, which enhances the underlying analysis. In this manner, important features, such as commonalities and anomalies, can be highlighted, analyzed and disseminated.

In order to make appropriate choices in terms of the methodology used to analyze and visualize the data, it is important to understand the different types of data, as these data types often directly impact the choices for visual representation and analysis. As such, one can refer to the *scales of measure* developed by Stevens [125] in which four different types of scales were defined to describe data measurements: nominal, ordinal, interval and ratio. Each type of scale corresponds to a different means of describing data.

1.1.1   NOMINAL DATA

Nominal data is the category of data in which each data element is defined by a label. Here, the categories have no order. Instead, the data takes on non-numeric values, and the observations belonging to the nominal data class can be assigned a code in the form of a number. The use of numerals as names for classes is simply a naming convention, and it can be replaced by other conventions as well. For example, in Table 1.1, the columns of 'Name' and 'Team' represent nominal data. Here, one can check equivalence and see if one player has the same name as another or if players share a similar team; however, one cannot provide details on what an 'average name' would be amongst the top twenty baseball players. Yet, if one chooses to apply some sort of ranking to the nominal data (sorting the names alphabetically, by length, by some measure of the team's perennial success), then the data begins to fall into the class of ordinal data.

1.1.2   ORDINAL DATA

Ordinal data has a specified rank ordering, but there is no specified degree of distance between two measured items. In this data type, the number code assigned to the observation implies an ordering; in Table 1.1, one could assign a number to the players names, with 1 being assigned to C. Gonzalez, 2 being assigned to J. Votto, etc. This assignment represents their relationship to the position of their batting average; however, the relative distance between player 1 and 2 would be undefined in this scale, due to the fact that the distance between player 1 and player 2 may be greater than the distance between player 3 and player 4. Furthermore, the assignment of labels in the ordinal data scale need not be numeric. Ordinal data may be assigned to named scales with implicit order, such as: 'very easy', 'easy', 'average', 'hard', or 'very hard'.

1.1.3   INTERVAL DATA

Interval data is data with specified distances between levels of an attribute. In this data type, the distance between two data pairs are able to be meaningfully compared as opposed to the distance between nominal and ordinal data. The most common example of interval data is the Fahrenheit scale. In this example, the distance between degrees is well defined; however, the ratio between the numbers are not meaningful (for example, one would not describe a summer day as twice as warm as a winter day). On the other hand, the ratio between differences are meaningful on the interval scale.

1.1.4   RATIO DATA

Ratio data is data which can be informally described as having a zero point that indicates the absence of the item being measured. Most measurements are done on ratio scales, including things like time and height. With respect to statistical descriptions, ratio data allows the use of all mathematical operations, thus allowing for descriptors ranging from the mean of the data to standardized moments.

While the use of these data classifications are widely adopted, they are not universally accepted. The introduction of these data types serves merely as an introductory means of describing the types and relationships of data common amongst structured data. The discussion of such data types provides a means for defining data structure in terms of the representation and relationships. By utilizing the underlying structure of the data and dealing with some notion of data types, one can begin to formulate ways to apply analyses and visually represent data. Ware [138] notes that only three of these levels of measurements are widely used and that the typical basic data classes most often considered in visualization are more influenced by the demands of computer programming.

CHAPTER 2

Color Schemes

One of the key components of visually representing data is choosing the appropriate color scale. The choice of the color scale is a complicated design choice that depends both on the data type, the problem domain and the applied visual representation. However, there is no 'best' color scale. Rheingans [115] notes that a choice of an appropriate color scale is influenced by several factors including the characteristics of the data being analyzed, the questions the analysts wishes to answer about the data, and the internal preconceived biases of the data representation that an analyst has. While there may not be a so-called best color scale choice, there are a series of design principles and guidelines described by Trumbo [132] and Levkowitz and Herman [93], which can provide insight into what color scale choices will be appropriate for a given visual analysis task.

2.1 DESIGN PRINCIPLES FOR COLOR SCHEMES

The first principle emphasized by both Trumbo [132] and Levkowitz and Herman [93] is that of *order*. Given a univariate data type in which the variable may be continuous or discrete, the color scale that is chosen to map the univariate data to a given color must represent some perceived ordering. This principle is directly relatable to data types that fall in the categories of ordinal, interval and ratio. Thus, given a sequence of ordered data, if we have a mapping such that our ordered data values $D = \{d_1 \leq d_2 \leq ... \leq d_n\}$ map to a set of colors $C = \{c_1, c_2, ..., c_n\}$, then the order found in D should be preserved in the color mapping C. Empirically, this means that c_1 should be perceived as be less than c_2, c_2 as less than c_3, and so forth, such that $C = \{c_1 \leq c_2 \leq ... \leq c_n\}$.

The second principle emphasized by Trumbo [132] deals with the *separation* of colors in a color mapping. What is meant by separation is that important differences between ranges of the variable should be represented by colors that can be perceived as being different. That is to say that within the color mapping C, c_i and c_j should be perceived as different for any $i \neq j$. Levkowitz and Herman [93] expands on this principle stating that not only should the colors be perceived as being different, but the distance between the colors should be perceived as equal.

Moreland [107] further summarizes a series of color map requirements. He notes that the color map should be aesthetically pleasing, contain a maximal perceptual resolution and that the ordering of colors should be intuitive for all users. Moreland [107] notes that while the requirement that a color map be aesthetically pleasing may have little to do with its effectiveness as a visualization, the fact that an end user can be discouraged by poor choices of aesthetics is an important consideration.

2.2   UNIVARIATE COLOR SCHEMES

Given the set of design principles, it is imperative to understand the pros and cons of color schemes in order to apply an appropriate mapping. A variety of univariate color maps already exist in the literature. The most commonly used scales include the rainbow and grayscale color maps. Figure 2.1 provides a set of examples of commonly used scales, and each scale has its own strengths and weakness with regards to the underlying analysis questions being asked of the data.

Figure 2.1: Sample univariate colormaps.

2.2.1   QUALITATIVE COLOR SCALES

The *rainbow color scale* is one of the most commonly used color maps in visualization. In fact, Borland and Taylor [15] find that the rainbow color map was used in 51% of all IEEE Visualization papers from 2001 to 2005. However, the rainbow color scale has been shown to be a poor color map in a large variety of problem domains. Rheingans [115] defines the rainbow color map as a *spectrum scale*, which is formed by holding the saturation and brightness constant and letting the hue vary through its entire range. This scale tends to follow the colors of the rainbow; however, the ordering of the hues is unintuitive. While this violates the ordering principle, the rainbow color map is not without its use. The rainbow color map and spectrum scale are also known as *qualitative* schemes in the cartography community, and a sample qualitative scale is shown in Figure 2.1. Work by Harrower and Brewer [65] utilize such schemes when working with nominal data types in which each data label or category can be separated into its own color. Due to the fact that nominal data has no implied ordering, there is no need to use a color map that conforms to the ordering principle. In fact, showing nominal data with an ordered color scheme could lead to misinterpretations about the underlying data itself.

2.2.2   SEQUENTIAL COLOR SCALES

Perhaps the simplest color scale is the *grayscale* color map in which the value of a single scalar variable is mapped to its brightness. This type of color scale can also be extended to a more general notion of the *sequential* color map described by [65]. An example of both grayscale and sequential color maps can be found in Figure 2.1. In the sequential color map, ordered data (such as ordinal, interval and ratio data) can be represented. These maps are designed with both the principle of ordering and separation in mind. Dark colors are typically used to represent higher ranges of the data, with light colors representing lower ranges of the data. However, this rule of 'higher equates to darker' is not essential; the crucial factor is simply relating the univariate data with a lightness sequence. The biggest advantage of this type of scale is that it is intuitive; however, its weakness lies in the limited number of distinguishable values that can be represented.

2.2.3   DIVERGENTCOLOR SCALES

Finally, the *divergent* color map provides a means for variable comparison. Rheingans [115] defines the divergent color map as a *double-ended scale*, which is created when two sequential color scales are pasted together as some shared end point (typical the lighter range end point). This type of scale is best suited for the ratio data types in which there is some meaningful zero midpoint. What this scale lacks is a natural ordering of colors. As such, careful choice must be taken when choosing a high and low end representation for the scale. Often, this is done with the concept of 'cool' colors and 'warm' colors as defined by Hardin and Maffi [Hardin and Maffi], where red and yellow colors are considered warm and blues are considered cool. One can think of divergent schemes, as back-to-back sequential schemes, which are centered around a critical value (often zero).

2.3   MUTLIVARIATE COLOR SCHEMES

While the most common color mapping is done with univariate data, schemes do exist for multivariate data. Figure 2.2 shows both a bivariate and trivariate color mapping scheme. Such techniques are most often used in statistical mapping to display the relationship between multiple variables on a single map.

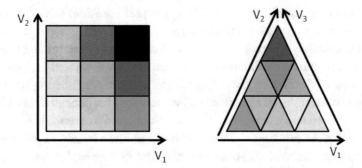

Figure 2.2: Sample multivariate colormaps. (Left) A bivariate colormap. (Right) A trivariate colormap.

The bivariate color map is an array of colors which can be created by crossing two univariate retinal variables. The underlying concept behind the bivariate color map is that the underlying bivariate data will be partitioned into a small number of classes, and each class will be assigned a color. The result is called an *overlay* scheme (as defined by Trumbo [132]). Along with the principles of order and separation, Trumbo [132] also details two additional principles for bivariate maps. The first principle is that of *rows and columns*. In order to preserve the univariate information within one dimension of the bivariate map, then the levels of the variables should not interact in a way that would obscure one another. The second principle is that of the *diagonal*. This principle states that if the goal is to visualize the association between variables, then the scheme of the bivariate map should resolve itself into three elements: the upper triangle, the lower triangle and the diagonal. As with the other principles, the principles of rows and columns and diagonals need only apply when dealing with ordinal, interval and ratio data.

However, higher-dimensional color schemes have been met with much criticism. Users note that the schemes often lack an intuitive progression from low to high, thus violating the principle of order and separation. This problem is further compounded when moving to higher dimensional schemes (see the three-dimensional color map of Figure 2.2).

Other methods for higher dimensional mapping schemes are often supported using texture overlays as a means of mapping higher level variable characteristics, such as in MacEachren [97]. Pham [113] attempted to utilize splines for generating univariate, bivariate and trivariate color schemes, and more recent work by Miller [102] applies an attribute block, which is a $k \times k$ array where each block in the array is colored by some attribute. While methods for higher dimensional color mapping schemes have been applied, many users still find such complex schemes to be too difficult to interpret.

2.4   CHOOSING A COLOR SCHEME

Given the variety of color schemes available and the underlying design principles in their creation, it is important to also look at the broader importance of the color selection choice. Color selection is a single component of the visualization, and Rheingans [115] provides several concepts to consider when attempting to design an effective visualization.

If the goal is to simply compare categories of data (looking at states that voted republican or democrat), it is best to choose qualitative color scales. If there is an underlying order in the data being analyzed, sequential scales are more appropriate, and when comparing about a critical value, the use of divergent color scales can be effective. Furthermore, the choice of colors will perceptually enhance portions of the data, drawing attention to features and locations. If the chosen color map is applied haphazardly, the resultant visualization could unintentionally emphasize non-important features, mask important features, and mislead the analyst in their exploration.

Depending on the problem domain being analyzed, color schemes exist which were designed to map certain colors to certain phenomena. This is true in weather maps and temperature scales, and other mappings may be more appropriate due to the underlying domain assumptions even if these maps may be less optimal in terms of the design principles. One type of color map to consider with respect to the audience is the stop light metaphor in which alerts are mapped to green, yellow and red with green being the lease severe alert and red being the most severe. Color choices for two-dimensional visualizations may not be optimal for three-dimensional visualizations. For example, grayscale maps can interfere with the interpretation of shading on three-dimensional objects.

Given the manner in which the choice of a color mapping influences the resultant visual analysis, it is imperative that appropriate design consideration is taken when creating visual analytic displays. Color schemes should be chosen with respect to the underlying data being analyzed, the type of analysis being performed, and the preconceived notions that the analyst will have in exploring the data. Furthermore, it is also important to realize that quantitative information about the data is not the only important issue in the analytic process. Often times, the end user is searching for changes in the data, thus, depending on the questions being asked of the data it could be best to create sharp perceptual changes locally within the data. As the user searches through the data, we can provide tools for interaction in which the user may vary the graphical characteristics of the visualization as part of the exploration and analysis process.

The recent Color Lens work by Elmqvist [52] is one example of applying such interactive techniques to color mapping. In this work, the range of data mapped to a particular color is dynamically modified, based on the user exploration. While the addition of interactive properties for modifying and exploring color maps can greatly benefit the visual analytics process, users must be aware of how changing color schemes can greatly impact the resultant perception of the information.

CHAPTER 3

Data Preconditioning

While the choice of color is a dominant component in creating an appropriate visual representation, the underlying distribution of the data structure is also a key factor in determining not only the appropriate visual display parameters but also the underlying statistical analyses. Under the assumptions of data normality, choices of data grouping and axes scaling for visual analysis have been well studied(e.g., [19, 29, 30, 31, 36, 70, 103, 140]). However, real world data often fails to meet any approximation of a normality assumption. One of the most effective ways of transforming data to a suitable approximation of normality is to utilize a power transformation. The power transformation was introduced by Tukey [135, 136] and further discussed as a means of visualizing data by Cleveland [35].

This concept of pre-conditioning data (utilizing a power transformation as an initial step) for analysis and visualization is well established within the statistical community and is employed as part of statistical modeling and analysis. However, within the visualization community, the application of appropriate power transformations for data visualizations is largely ignored in favor of interactive explorations (e.g., [68]) or default applications of logarithmic or square root transforms (e.g., [85]). Yet, transformation is a critical tool for data visualization as it can substantially simplify the structure of a data set.

Traditionally, statisticians have applied data transformations to reduce the effects of random noise, skewness, monotone spread, etc. [35], all of which can affect the resulting data visualizations. For example, reducing random noise can help show global trends in the data, changing the range of values can help fit the data on displays with small screens, and reducing the variance can help improve comparative analysis between multiple series of data. In approximately normal data, methods of data fitting and probabilistic inference are typically simple and often more powerful. Furthermore, the description of the data is less complex, leading to a better understanding of the data itself. As such, by choosing an appropriate power transformation, data can often be transformed to a normal approximation, lending itself to more powerful visual and analytical methods.

The choice of an appropriate power parameter is the most important aspect of the application of the power transform. Power transformations help to achieve approximate symmetry, stabilize variance across multiple distributions, promote a straight line relationship between variables and simplify the structure of a two-way or higher-dimensional table [16, 41, 126, 135]. The power transformation [135] is a class of rank-preserving data transformations parameterized by λ (the power) defined as:

$$x^{(\lambda)} = \begin{cases} x^{\lambda} & (\lambda \neq 0) \\ \log(x) & (\lambda = 0) \end{cases} \tag{3.1}$$

where x is the observed or recorded data.

Under this transformation, for $\lambda = 1$, the data remains untransformed, for $\lambda = -1$, the data is inverted, etc. For data skewed towards large values, powers in the range of $[-1,1]$ are generally explored. Powers above 1 are not typically used if the data has large positive values because it increases the skewness. It is also commonly observed that as the power is reduced from 1 to -1, the data is transformed until it is nearly symmetric, and upon further reduction, it becomes asymmetric again [35]. This is important for visualization as skewed data tends to result in overly large graphs to represent the full dynamic range or graphs where outliers are visible, but data near the mean of the distribution are grouped together.

Statistically, the goal is to find a suitable power for the most appropriate transformation such that the variance in the data is stabilized. Such a value helps in conditioning the data, enabling easier data analysis in subsequent stages. At the same time, it also leads to desirable changes in the data that helps to improve visualizations in 1D and 2D. Traditionally, an appropriate power for the power transformation is chosen through trial and error, by plotting the mean of each data series versus its standard deviation for different powers from a finite set of possible powers determined empirically. Typical choices that are used by statisticians for the power are $\{-1, -\frac{1}{2}, -\frac{1}{4}, 0, \frac{1}{4}, \frac{1}{2}, 1\}$ since they provide a representative collection of the power transformation [35]. Based on this statistical observation, an appropriate power can be chosen to make the distribution symmetric. Statistically, this means that the data distribution is rid of spread variation, thus leaving one with only location variations, which are easier to model. In many cases, the power chosen using the above method also brings the data closer to normality, which is always a desired effect in data modeling. While this method of interactively selecting the power transformation provides more control over the choice of the power for each dataset, it is cumbersome and may not always result in the best possible power as one cannot examine all the possible choices.

One alternative to the trial-and-error approach would be to utilize the Box-Cox family of power transformations [16]. The transformation, introduced by Box and Cox [16], is a particular family of power transformations with advantageous properties such as conversion of data to an approximately normal distribution and stabilization of variance. Given a vector of n observations $x = \{x_1, ..., x_n\}$, the data is transformed using the Box-Cox transformation given by:

$$x^{(\lambda)} = \begin{cases} \frac{x^\lambda - 1}{\lambda} & (\lambda \neq 0) \\ \ln(x) & (\lambda = 0) \end{cases} \tag{3.2}$$

where x is the vector of observed or recorded data and the parameter λ is the power. Note that the above formula is defined only for positive data. However, any non-positive data can be converted to this form by adding a constant.

Given this initial transformation, Box and Cox [16] then assumed that for some unknown λ, the transformed observations $x_i^{(\lambda)}$ $(i = 1, ..., n)$ are independently, normally distributed with constant variance σ^2 and with expectations,

$$E\{x^{(\lambda)}\} = a\theta, \tag{3.3}$$

where a is a known matrix and θ is a vector of unknown parameters associated with the transformed observations. The likelihood, in relation to the original observations, x, is obtained by multiplying the normal density by the Jacobian of the transformation, thus

$$\frac{1}{(2\pi)^{\frac{n}{2}}\sigma^n} \exp\left\{ -\frac{(x^{(\lambda)} - a\theta)'(x^{(\lambda)} - a\theta)}{2\sigma^2} \right\} J(\lambda; x), \tag{3.4}$$

where

$$J(\lambda; x) = \prod_{i=1}^{n} \left| \frac{dx_i^{(\lambda)}}{dx_i} \right|.$$

One can then maximize the logarithm of the likelihood function and readers are referred to the work of Box and Cox [16] for details and derivations. The final derivation for the maximum likelihood estimator yields,

$$L_{\max}(\lambda) = -\frac{1}{2} \log S(\lambda; y)/n, \tag{3.5}$$

where

$$S(\lambda; y) = y^{(\lambda)'} a_r y^{(\lambda)}, \tag{3.6}$$

$$a_r = I - a(a'a)^{-1}a', \tag{3.7}$$

and

$$y^{(\lambda)} = x^{(\lambda)}/J^{\frac{1}{n}}. \tag{3.8}$$

Finally, λ can be maximized by taking the derivative of $L_{\max}$ with respect to λ and finding the critical points. In the special case of the one parameter power transformation, $x^{(\lambda)} = (x^\lambda - 1)/\lambda$,

$$\frac{d}{d\lambda} L_{\max}(\lambda) = -m \frac{x^{(\lambda)'} a_r u^{(\lambda)}}{x^{(\lambda)'} a_r x^{(\lambda)}} + \frac{n}{\lambda} + \sum \log x_i. \tag{3.9}$$

where, $u^{(\lambda)}$ is the vector of components $\{\lambda^{-1} x_i^\lambda \log x_i\}$.

Once an appropriate power transformation is chosen, the data is transformed, which, in turn, means the axis on which the data is plotted is also transformed. Such transformations are of key significance when data is skewed, and the guidelines from a statistical visualization viewpoint [35] describe the benefits of statistically pre-conditioning skewed data in practice. Skewed data is data in which the majority of the samples lie near the median values with outliers stretching the data domain to large (or small) values, thus increasing the range needed for a given display axis. The plotting of this skewed data compresses values into small regions of the graph, resulting in a lower fidelity of visual assessment of the data [35]. One option to improve data assessment would be to remove the outliers and focus on the range of data near the median, requiring an interactive technique such as zooming, or users may select the data they are interested in (by brushing) to create a new plot that focuses on the subset of interest. Another option is to apply an appropriate choice of power transformation as a pre-processing step and use this power transformation to transform the axis. This transformation reduces some of the need for interaction and massages the data into a form that is statistically more suitable for advanced analytical techniques.

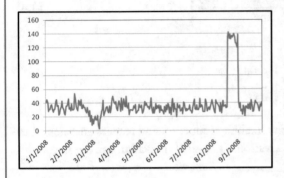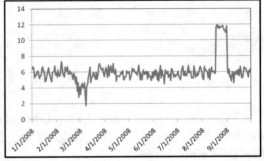

Figure 3.1: Applying the power transformation when plotting time series data. The images shown represent a time series plot of simulated patient counts using data generated by Maciejewski et al. [99]. (Left) The untransformed data. (Right) The transformed data.

The effects of utilizing such transformations are illustrated in Figure 3.1. Figure 3.1 shows a plot of total patient visits to a hospital, and one can clearly see that most of the data is compressed to the bottom of the graph as it needs to accommodate both high and low values. However, the Box-Cox transformation can be used to find a suitable power to transform the data that better utilizes the space, allowing us to simultaneously see the spike as well as the detail in the previously compressed region. For example, in the transformed plot (Figure 3.1 - Right), the dip in the graph near 3/1/2008, and its corresponding fluctuations are more easily explored when compared to the original untransformed plot (Figure 3.1 - Left).

While the power transformation is a powerful tool, there are a limited number of cases in which it is not appropriate to use, particularly, in cases where the power is outside the range of [-1,1]. In these cases, the data may be overly exaggerated or inverted, depending on the sign of the power. Therefore, the application of this procedure should be limited to data plots that contain at least one skewed plot that is significantly non-normal. Normality of the given data plot can be measured automatically by computing the correlation coefficient of its normal probability plot and thresholding the coefficient value. Moreover, the automatically computed power should be checked if it is in the range [-1,1] before application of this procedure. Furthermore, working in the transformed space may be difficult for users as the axis labels are now related to the data only through a relatively complex mathematical formula. As such, one needs to be judicious as to when to use such transformations, or if they are better suited as a preprocessing step in the underlying analytical analysis.

Previous work has looked at utilizing power transformations for axis transformations. For example, Cook and Weisberg's Arc system [37] has utilized interactive interfaces in which the user can drag sliders to change the Box-Cox transformation or simply click a button to set the transformation to the log-likelihood-maximizing value. However, many current visual analysis tools still fail to consider the underlying data distribution and instead rely on user intuition. For example, Tableau incorporates frequency plots and histograms and groups the data into bins of equal width; however, the frequency plots and binning used often results in suboptimal visual displays for comparison and analysis, and users often will resort to interactive techniques to zoom into the data or manually adjusting bin sizes to remove the effects of outliers. Such procedures can become tedious and often inaccurate, especially when skewed data is involved. Thus, there is a need for the continued exploration and application of power transformations for enhancing both the visual representation and underlying analytical processes.

CHAPTER 4

Visual Representations and Analysis

Visual representations provide the means to visually analyze the data; however, visual analytics is more then just a visual analysis. Shneiderman [121] coined the mantra of "Overview, zoom and filter, details on demand." This idea of providing overviews of the data and then details on demand has been readily adopted in the visualization community; however, the notion of visual analytics extends this idea. Here, the overview step is now replaced by an analysis step which is used to direct the overview to the important aspects of the data. The zoom and filter step then allows for user interaction, with a continual loop of returning to the analysis stage for further refinement. As such, the choice of both the visual representation and the underlying analysis are greatly intertwined. In the previous sections, the connections between the data and the visual representations were explored in terms of color, while techniques were discussed with regards to preprocessing the data using power transformations to make the data more manageable in terms of both the visualization and analysis. In this section, the linkages between the visual representation and the analysis are further discussed.

4.1 HISTOGRAMS

In exploring and analyzing data, perhaps the most common initial exploration of the data is done through the use of the histogram, which was first introduced by Pearson [111]. The histogram provides a visual summary of a single variable distribution within a dataset consisting of frequency counts of the data represented as rectangles over discrete intervals (called *classes* or *bins*). The height of each rectangle corresponds to the frequency density of the bin.

4.1.1 DETERMINING BIN WIDTHS

According to Wilkinson [142], the histogram is probably the most widely used visual representation and first look analysis tool; however, it is arguably one of the more difficult ones to compute. The main concern in creating a histogram lies within the choice of the number of bins. However, there is no optimal number of bins. Different numbers of bins and different bin sizes can each reveal different insights into the data. Figure 4.1 shows four histograms of the player batting averages provided in Table 1.1, created using different choices for either the number of bins or the bin width.

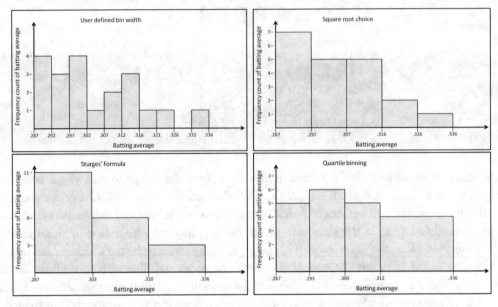

Figure 4.1: A set of histograms, created using different binning rules, showing the distribution of the batting average statistic provided in Table 1.1. (Top - Left) The number of bins (k) is arbitrarily chosen by the user. (Top - Right) The number of bins (k) is defined by the square root choice, Equation 4.2. (Bottom - Left) The number of bins (k) is defined by Sturges' formula, Equation 4.3. (Bottom - Right) The number of bins (k) is defined by the interquartile ranges of the data.

In Figure 4.1 (Top-Left), the histogram was created using an arbitrary choice of ten bins. The bin width, h, can then be solved for in the equation:

$$k = \left\lceil \frac{\max(x) - \min(x)}{h} \right\rceil. \tag{4.1}$$

However, it is atypical that the user will actually define the number of bins. Instead, most histogram creation tools utilize equal bin width rules, defining the number of bins with a square root relationship to the size of the data, n,

$$k = \left\lceil \sqrt{n} \right\rceil. \tag{4.2}$$

This is the default in many common histogram packages including Excel. The application of the square root binning method to the batting average data of Table 1.1 is shown in Figure 4.1 (Top-Right).

Another common bin selection includes Sturges' choice [127]:

$$k = \left\lceil \log_2(n) + 1 \right\rceil. \tag{4.3}$$

The application of Sturges' binning method to the batting average data of Table 1.1 is shown in Figure 4.1 (Bottom-Left). Both the square root choice and Sturges' choice have an implicit assumption of a Normally distributed dataset.

Other methods have focused on incorporating smoothing criteria such as the integrated mean square error. For a Normal distribution, Scott [119] proved that the optimal integrated mean square error bin width is given by

$$h = \frac{3.5\sigma}{n^{\frac{1}{3}}}.$$ (4.4)

While methods such as [127] and [119] focus on equal bin width histograms, one could also choose to make them all have approximately the same area, Figure 4.1 (Bottom-Right). In this case, one could utilize a quantile binning procedure in which approximately the same number of samples fall within each bin. Quantiles are points taken at regular intervals from the cumulative distribution function, thus dividing ordered data into q subsets of approximately equal size. A variety of methods can be used for estimating quantiles. Figure 4.1 (Bottom-Right) illustrates the application of quantile binning to the batting average data of Table 1.1.

Here, it should be noted that most statistical packages use Sturges' rule [127] or some extension of it when determining the number of bins. However, in the construction of the approximation, Sturges considered an idealised frequency in which the distribution will approach the shape of a normal density. Thus, if the data is not normal, the number of bins chosen will need to be revised. One solution is to utilize some sort of data preconditioning technique as describe in Chapter 3. However, for moderately sized n (approximately $n < 200$), Sturges' rule will produce reasonable histograms.

As such, it is important to be aware of the strengths and weakness of the histogram binning choices. The equal-width histogram tends to over smooth data in regions of high density and often fails to identify sharp peaks. The equal-area histogram over smooths in regions of low density. With small bins, the histogram is able to model fine details within the data; however, the estimation of the density is too rough as there will be many local maxima and minima. With large bins, the density becomes too smoothed, and one can lose properties of the data. More modern approaches have tried to overcome these issues by creating approaches that attempt to reconcile these two defects; for example, work by Denby and Mallows utilize the asymptotics of the Integrated Mean Square Error to create histogram bin recommendations [42].

Visual analytic packages often allow users to interactively adjust bin widths or choose the number of bins for classification. While such interactive exploration can enhance the underlying analytical process, the analyst may also choose bins that obfuscate the data. Thus, it is important to realize that the histogram is an aggregation tool. With too few bins, important patterns may be hidden, and with too many bins, the underlying data noise may clutter these patterns. Furthermore, as the data dimensionality increases and analysts need to compare data distributions and look for patterns over a multitude of variables, it is often the case that the initial data representation will be the leading guide in where they search. As such, it is imperative that appropriate choices for the underlying data analysis (in the case of the histogram analyzing, the appropriate choice for data binning) be made prior to the visualization of the data. That is not to say that the analysis and visual representation chosen in the first step should be the final state that the analyst observes; however, it should not deter the analytical process through poor choices in the initial exploration phase.

4.1.2 INCREASING THE DIMENSIONALITY OF A HISTOGRAM

As previously mentioned, the analytical process often involves the use of multi-scale, multi-source data sets. These data sets are rife with linked, multivariate information, and many tools and techniques have been created to analyze and visualize these linkages. While the histogram is a powerful first look tool for data visualization, many datasets are typically high-dimensional and require a large number of histograms to represent the various correlations.

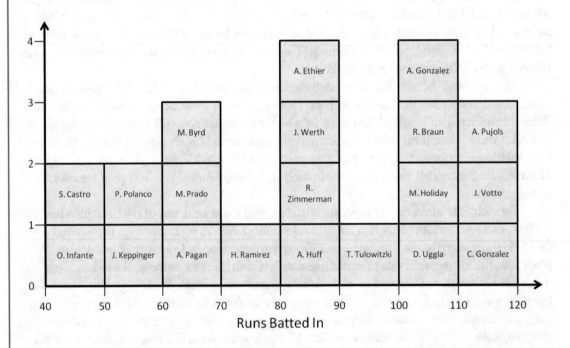

Figure 4.2: A dot plot histogram of the distribution of runs batted in of the top 20 Major League Baseball players (by batting average) where each dot contains the label of the player that makes up the dot. Data for this graph is provided in Table 1.1.

Wilkinson [141] proposes the use of dot plots to depict a higher level of detail within distributions. The dot plot takes a set of univariate observations $x_1, x_2, ..., x_n$ and starting with the smallest data value, draws a dot on the graph. The size of the dot provides the visual appearance of the plot, and the choice of dot size is analogous to the choice of the histogram bin width. Once the first dot is drawn, the next smallest data value is taken, and it is either stacked on the first dot (if their radius would overlap) or it is placed as a new starting element. Once all the elements are placed, a sort of histogram is revealed. The benefit of such a plot over the histogram is that it can reveal local features because the dots in the dot plot can be labeled. Figure 4.2 shows a variation of the dot plot in which the dots are now square boxes, and each box contains a further label of the data element. Here, the distribution of runs batted in by the top twenty Major League baseball players (by batting average) has been plotted. By labeling the 'dots' in the plot, the analyst can see not only the distribution of data within a histogram but also detailed information about the elements within each bin. However, such methods would only work for reasonably sized datasets. As datasets grow larger, each dot would become an aggregation of several dots, and individual labeling would be replaced by group labeling and result in less detailed information. Recently, Dang et al. [40] extended Wilkinson's work on dot plots to stacked dot plots in multi-dimensions and other statistical graphics.

Another common means of showing more dimensionality with histograms is through stacking bars. A stacked histogram is used to compare parts of a whole; that is, each bin in the histogram is divided into a category, which the divisions being used to represent the contribution to the total that a given category will make. Figure 4.3 illustrates the use of a stacked bar chart in analyzing sales data for a company across three separate stores. Here, each store is represented by a different color. The stacked histogram provides insight into the trends of the overall dataset and the proportion of each category within each histogram bin. However, the stacked histogram visualization makes it difficult to do inter-category comparison or trend analysis as each block can be at a different height.

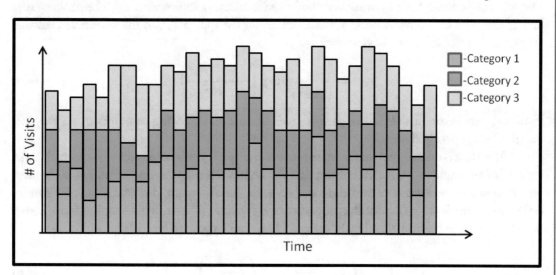

Figure 4.3: A stacked histogram showing total sales from three different stores over a period of time.

Other methods of enhancing the dimensionality of histograms would be through highlighting in which users can interactively select bars in the histogram and details of the aggregate data are shown to the user as a means of providing more information. Such interactive techniques are common in most visual analytics systems and have been used in a variety of data exploration problems. Overall, the histogram is a powerful first look data analysis tool that allows analysts to understand the distribution of their data, search for outliers and form hypotheses on what the shape of the data distribution means. However, the visualization of the histogram is highly subject to choices in bin widths. Furthermore, as the number of dimensions in the data increases, the number of bins needed to represent the data increases, growing exponentially. Thus, histograms are only appropriate for data sets with few dimensions.

4.2   KERNEL DENSITY ESTIMATION

While histograms are a powerful tool for data visualization and analysis, it was noted that one of the key problems in creating histograms are choosing appropriate bin widths. An alternative to using the histogram as a means for exploring data distribution is the application of kernel density estimation [122].

The goal of kernel density estimation is the same of that of a histogram, to provide an approximation of the underlying probability distribution of a data set from a set of observed data. One approach to density estimation is *parametric* in which one assumes that the underlying distribution of a family is known and the parameters of that distribution are then estimated from the observed samples. An example of this would be given a set of observed samples $X = x_1, x_2, ...x_n$; it is assumed that this data is drawn from a normal distribution with means μ and variance σ^2. The parameters μ and σ^2 can then be estimated using a maximum likelihood approach, and the univariate Gaussian (or Normal) distribution can be described as

$$p(x) = \frac{1}{(2\pi\sigma^2)^{\frac{1}{2}}} \exp\left\{-\frac{(x-\mu)^2}{2\sigma^2}\right\} \tag{4.5}$$

However, parametric models fail to perform well when the underlying assumptions on the data distribution are poorly chosen.

As such, it is often useful to rely on *non-parametric* density estimation techniques, which make no *a priori* assumptions on the underlying data distribution. Kernel density estimation is one such means of non-parametric density estimation. As with the histogram, one of the uses for density estimation is the investigation of the properties of a given set of data. In its simplest form, kernel density estimation (as formulated by Silverman [122]) can be written as:

$$\hat{f}(\mathbf{x}) = \frac{1}{Nh} \sum_{i=1}^{N} K\left(\frac{\mathbf{x} - X_i}{h}\right) \tag{4.6}$$

where h is the *smoothing parameter*, N is the number of samples and K is the kernel estimator.

Just as the histogram is a series of boxes with a given width, kernel density estimation relies on a kernel estimator which determines a window with that is analogous to the concept of the bin width in creating histograms. The kernel function determines the shape of the window, with the most common kernel being the Gaussian kernel, which is of the same form as Equation 4.5. Figure 4.4 illustrates the application of kernel density estimation to a sample of four univariate points using a Gaussian kernel. Here, one can see the effects of varying the bandwidth, h. As h is reduced, the function tends towards a series of spikes, and as h becomes large, all detail is obscured. Note that the sample size in this example is $N = 4$. It is not usually appropriate to construct density estimates from such small samples, and this example is used only as a means of illustrating the application of kernel density estimation.

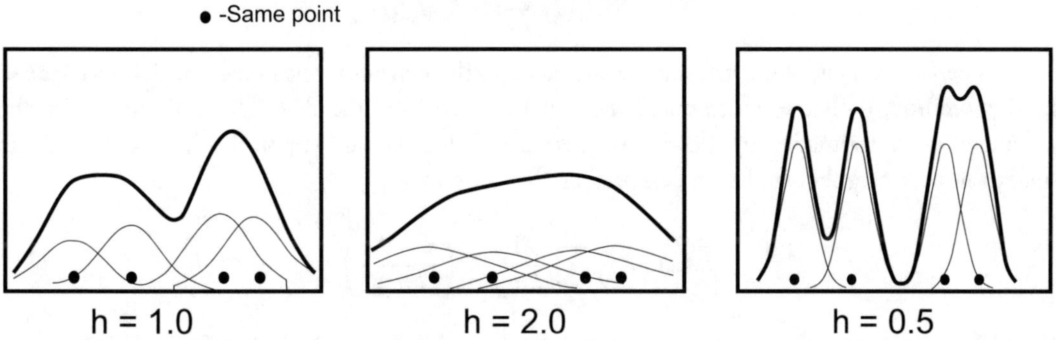

Figure 4.4: Kernel estimates showing individual kernels to illustrate the effects of changing the bandwidth parameter.

Apart from the histogram, the kernel estimator is one of the most commonly used estimators. In density estimation (as with histograms), the choice of the bandwidth parameter (or smoothing parameter) is the most crucial step. As in Sturges' rule, one can make underlying assumptions about the data and choose a bandwidth parameter that would be related to the variance of the data. Such a choice works well if the underlying distribution is actually normally distributed; however, it tends to over smooth in the case that the data is not normally distributed. Much like in the histogram case, one could apply various preconditioning methods to the data to reduce the impact of some of these issues. Silverman [122] suggests that better results can be obtained by using a measure of the interquartile range of the data; however, in the case of bimodal distributions, an adaptive method of the spread would be better utilized.

Adaptive methods for kernel density estimation are based on the *nearest neighbor* class of density estimators. These nearest neighbor estimators attempt to adapt the bandwidth based on the 'local' density of the data. The degree of smoothing is chosen by an integer, k, which represents the k-th nearest neighbor. A typical choice is for $k \approx \sqrt{N}$. Then, the distance from the current sample to its k-th nearest neighbor is calculated such that the distance $d(x, y)$ between two points in the sample is $|x_1 - x_2|$. For all $d(\alpha)$, the distance from x_i to all other sample points are calculated and arranged in ascending order. The k-th nearest neighbor density estimate is then defined by

$$\hat{f}(\mathbf{x}) = \frac{k}{2Nd_k(x)}. \tag{4.7}$$

Thus, in distributions with long tails, the distance $d_k(x)$ will be larger than in the main part of the distribution, thereby reducing the problem of under smoothing the tail. This estimation is then generalized to a kernel smoother in the form of

$$\hat{f}(\mathbf{x}) = \frac{1}{Nd_k(x)} \sum_{i=1}^{N} K\left(\frac{x - X_i}{d_k(\alpha)}\right) \tag{4.8}$$

The nearest neighbor approach is also related to the more complicated *variable kernel method*. In this method, both the width and height of the kernel are variable. This estimate scales the parameter of the estimation by allowing the kernel scale to vary based upon the distance from X_i to the kth nearest neighbor in the set comprising $N - 1$ points.

$$\hat{f}_h(\mathbf{x}) = \frac{1}{N} \sum_{i=1}^{N} \frac{1}{hd_{i,k}} K\left(\frac{\mathbf{x} - X_i}{hd_{i,k}}\right) \tag{4.9}$$

Here, the window width of the kernel placed on the point X_i is proportional to $d_{i,k}$ (where $d_{i,k}$ is the distance from the ith sample to the kth nearest neighbor in the set comprising the other $N - 1$ samples). Thus, data points in regions where the data is sparse will have flatter kernels as the smoothing parameter, h, will be proportional to the distance to the nearest neighbor.

While the kernel density estimation techniques, discussed up to this point, focus on univariate samples, one of the most important applications of density estimation is the analysis of multivariate data. The multivariate kernel density estimator can be defined as:

$$\hat{f}_h(\mathbf{x}) = \frac{1}{Nh^d} \sum_{i=1}^{N} K\left(\frac{\mathbf{x} - X_i}{h}\right) \tag{4.10}$$

where the kernel function, $K(x)$ is now a function defined for a d-dimensional x. K is a probability density function, defined by the user and is typically defined to be a Gaussian distribution in the form of Equation 4.5. The use of a single smoothing parameter h in Equation 4.10 implies that the kernels will be equally scaled in all directions. Depending on the application, it may be useful to transform h into a vector and scale the data, for example, if the spread of the data is skewed along one axis. Here, one could again consider preconditioning the data as was discussed in Chapter 3. This preconditioning will help avoid extreme differences of spread, and if applied correctly, there will typically be little reason to apply more complicated forms of the kernel density estimate involving a single smoothing parameter.

Figure 4.5 illustrates the application of the multivariate kernel density estimator applied to a distribution of sample emergency department visits across the state of Indiana. Each point

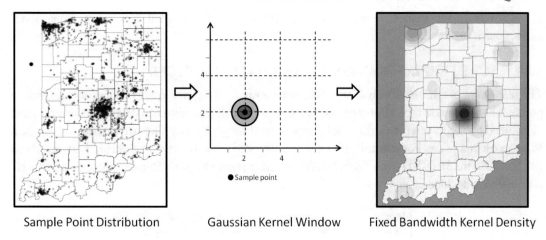

Sample Point Distribution   Gaussian Kernel Window   Fixed Bandwidth Kernel Density

Figure 4.5: The Kernel Density Estimation process applied to a distribution of emergency room visits in the state of Indiana.

utilizes a Gaussian kernel with a single smoothing parameter. The resultant density distribution is then provided, showing the multimodal distribution of patients across the state. In this manner, researchers can quickly gain an understanding of their dataset that would potentially be masked when simply plotting all given samples. The only potential drawback to the multivariate kernel method is the fact that in distributions with long tails, the main portion of the distributed density will be over smoothed. In order to overcome this issue, researchers have studied the use of the *adaptive kernel method*, in which the size of the kernel varies at each sample point. Details on this method and further thoughts on optimizing the kernel width can be found in Silverman's text [122].

Overall, one can apply kernel density estimation to a high-dimensional data set in order to approximate the overall distribution of the data. These distributions can be explored in lower dimensional projections and can be used as a means of providing insight to the data. Experts can also incorporate domain knowledge to define different bandwidths between dimensions. This analysis technique is able to provide a means for describing data based on a sample population.

4.3   MULTIVARIATE VISUALIZATION TECHNIQUES

While histograms and density estimation can be extended to view data sets of dimensions lager than two, as the dimensionality increases, the applicability of these tools as a means to visualize all aspects of the data set decreases. However, all visualization techniques rely on projecting data into a 1, 2 or 3-dimensional display. The visual ordering of these lower dimensional projections is a key issue that should be considered when utilizing many of these techniques. The arrangement of the order in which data and categories are displayed graphically can impact the analysis process either positively or negatively. In this section, details on exploration of high-dimensional data in low dimensional projections is explored.

4.3.1   SCATTERPLOTS AND SCATTERPLOT MATRICES

The scatterplot is a means of visualizing discrete data values along two axes as a collection of discrete points. These plots are typically employed as a means of analyzing bivariate data to see how they are distributed throughout a plane. A scatter plot is created simply by mapping two variables of the data to two axes in a plane. For example, if one were interested in seeing the relationship between Runs and Runs Batted In (RBI) from Table 1.1, one could create a plot with Runs on the x-axis and RBI on the y-axis, as done in Figure 4.6.

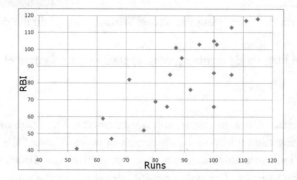

Figure 4.6: A scatterplot illustrating the relationship between Runs and Runs Batted In (RBI) from Table 1.1.

Scatterplots are used to help researchers understand the potential underlying correlations between variable [133]. These visualizations provide a quick means of assessing data distributions, clusters and outliers. In scatterplots, one can visually assess the correlation by looking for data trends. If the points tend to cluster in a band running from the lower left to the upper right, this is often indicative of a positive correlation between variables. Likewise, if the points tend to cluster in a band running from the upper left to the lower right, this can be indicative of a negative correlation. Immediately, one can begin exploring the data, observing outliers and trends. In Figure 4.6, the analyst can quickly find the outliers, where the player with the highest number of runs has an RBI near the median value. Furthermore, it seems that the data points to a positively sloping linear trend indicating that Runs and RBIs may be correlated (i.e., the players with a higher RBI have a higher amount of runs). However, Cleveland [35] notes though that putting a smooth curve through the data in the mind's eye is not a good method for assessing nonlinearity.

While scatterplots allow one to assess the bivariate relationship of data, the abundance of multivariate data has led to an extension of scatterplots in the form of a scatterplot matrix. A scatterplot matrix is (as the name says) a matrix of scatterplots where each column of the matrix contains the same x-axis and each row contains the same y-axis. Such a matrix is useful for visualizing how a dataset is distributed through multiple variables. By visualizing the bivariate distribution of all combinations of variables within a multivariate data set, one can quickly assess how clusters of points change shape between dimensions. Figure 4.7 shows a scatterplot matrix of the baseball statistics presented in Table 1.1.

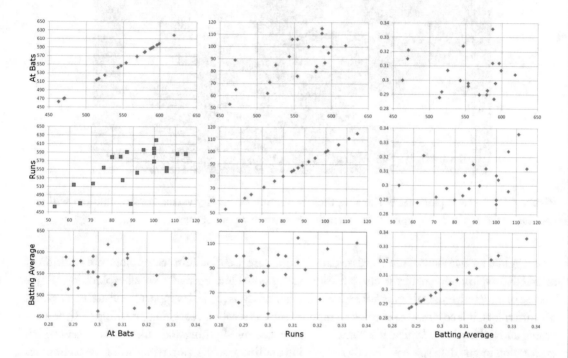

Figure 4.7: A scatterplot matrix showing the bivariate distributions across of combination of three variables from Table 1.1 (At Bats, Runs and Batting Average).

One important observation to note is that as the amount of data being plotted becomes larger, the visual clutter in a scatterplot can obscure the patterns. Here, one can utilize density estimation techniques discussed in Section 4.2 as a means of detecting or highlighting features that are not obvious from the scatterplot itself. For example, Figure 4.8 shows the application of density estimation to the scatterplot matrix representing the distribution of the density and density gradient within a volumetric dataset. In the leftmost image of Figure 4.8, each point in the scatterplot was given an opacity. In this manner, locations with more points would be brighter in the image. This allows users to assess the density of the points within the scatterplot; however, the application of density estimation provides more details as to the shape of the underlying distribution. Similar to the application of density estimation to scatterplots is the recent work of creating continuous scatterplots by Bachthaler and Weiskopf [12] where the authors propose a mathematical model of continuous scatterplots that considers an arbitrary density defined on an input field within a n-dimensional spatial grid.

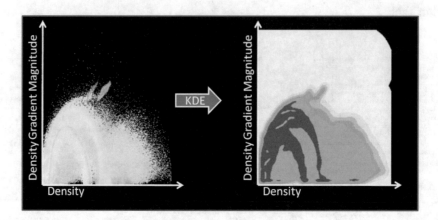

Figure 4.8: A scatterplot matrix showing the distribution of density versus density gradient within a volumetric dataset and the application of density estimation to this space.

While an overview of multivariate data relations are useful within scatterplots, the addition of interactive exploration enables users to further analyze the data. Formally, visual exploration [84] is the use of interactive visualization for exploratory data analysis [35]. To support the visual exploration of scatterplots, it is common to employ techniques such as scrolling, zooming, brushing [13], and focusing and linking [35]. Typical interactions include data highlighting and linked windows in which exploration in one data space is directly reflected in another visualization window. These techniques are typically utilized across the majority of visualization applications, often following Shneiderman's mantra [121] of "Overview first, zoom and filter, then details-on-demand." In brushing data, in scatterplot matrices, data points selected in one scatterplot window will highlight the same points in all other scatterplots within the matrix, allowing the user to see the relationships in different feature spaces of the data. More recent interactions with scatterplots include an extension called scatterdice, developed by Elmqvist et al. [51].

In the scatterdice technique, users interactively traverse the scatterplot matrix. However, the transition from one scatterplot to the next is animated in a view window in which the data is extruded from the current 2D space to include the new third dimension of data that is being transitioned to. Next, the space is rotated until the new plane is view aligned, and finally the new scatterplot plane is projected to the user. Figure 4.9 illustrates this transition from one scatterplot to another using the scatterdice method.

One final consideration in creating scatterplot matrices is the dimensional ordering of the matrix. Ideally, the ordering of the matrix could be done in such a way as to place dimensions that are similar next to each other in the matrix. Ankerst et al. [8] derive a set of similarity metrics for

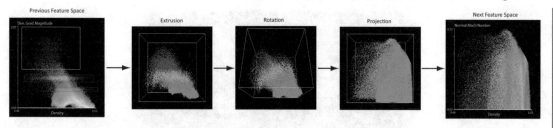

Figure 4.9: Animated transitions between feature spaces provides users with enhanced contextual information about data relationships. In this image, the user is navigating from the density vs. density gradient magnitude feature space to the density vs. normal Mach number feature space.

general data and note that the similarity of dimensions is an important prerequisite for finding the optimal arrangement of data features.

In creating scatterplots and scatterplot matrices for visualization and analysis, key factors include the amount of samples being plotted, the data dimensionality, and the relationship between dimensions. Not all dimensions of the data will need to be explored and visualized. Expert knowledge from the analyst can help guide the creation of scatterplots and scatterplot matrices to help reduce the amount of information that will be shown. Furthermore, by performing data preconditioning and analysis prior to creating scatterplots and scatterplot matrices, anomalies and commonalities within dimensions may be captured that can also be incorporated into the visualization through highlighting, axis scaling, and density estimation.

4.3.2 PARALLEL COORDINATE PLOTS

While scatterplots can provide an overview of relationships between multivariate data dimensions, they are limited in terms of screen space and the fact that they only present two variable relationships at a time. Another common method employed that can be used to show the relationship of variables over the entire n-dimensional space is the parallel coordinate plot [75].

Parallel coordinate plots show a set of points in an n-dimensional space. In the parallel coordinate plot, a set of parallel axes are drawn for each variable. Then, each data sample is represented by a line that connects the value of that samples variable attribute to all other attributes. As more variable attributes are added, more axes are added and more connections are made. An example parallel coordinate plot of the baseball data from Table 1.1 is shown in Figure 4.10. Here, on can begin to extrapolate the relationship between variables, as well as observe which players performances are comparable. For example, the player whose line is near the lower end of the first three variables (Castro) has a comparable batting average to the player with the highest number of at bats. Further, near the top of the plot, we find two players (Gonzalez and Pujols) who have nearly identical statistics in everything except Batting Average.

The value of such a plot is that certain geometrical properties in high dimensions transform into easily seen 2D patterns. For example, a set of points on a line in the n-dimensional space will

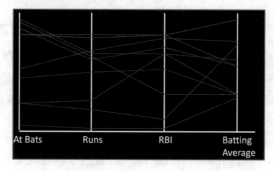

Figure 4.10: A parallel coordinate plot visualizing the top ten data samples from Table 1.1

transform to a set of polylines intersecting a $n - 1$ points within the parallel coordinate plot. Other known patterns that can be discerned are planes, curves, several smooth surfaces, convexity and non-orientability. Here, the order of axes is critical for finding features, and exploratory reordering needs to be applied, see Yang et al. [144] as an example of work that attempts to reorder the axes.

Furthermore, it should be noted that issues with cluttered displays arise as the amount of data being represented grows. Figure 4.10 utilizes only ten samples from Table 1.1, and already interaction techniques would be necessary to tease apart some of the data components. In scatterplots, the application of density estimation can still provide information with respect to the distribution of the underlying data. In a similar fashion, Heinrich and Weiskopf [71] present a continuous parallel coordinates method. In this work, the authors provide a 2D model that can be used to compute a density field for the parallel coordinate plot and the resultant visualization of this method.

4.3.3   PARALLEL SETS

While parallel coordinates and scatterplot work well for ordinal, interval and ratio data types, they do not work well for nominal data types where data falls to a discrete set of labels. Parallel sets [88] is a visualization application for categorical and nominal data. In a parallel set visualization, axes are positioned as in parallel coordinates; however, intersections at the axes are replaced with a set of boxes representing categories. The boxes are scaled corresponding to the frequency of the observations that fall into each category. In this way, the dimensional clutter can be reduced, and example of the parallel set application is shown.

4.3.4   ABSTRACT MULTIVARIATE VISUALIZATIONS

As in scatterplots, the projection of data to a lower plane in the parallel coordinate plot will result in a loss of information. In order to increase the number of dimensions being shown to a user at any given instance, other multivariate visualization techniques focus on the creation of glyphs. A glyph is a representation of a data variable that maps to graphical primitives or symbol [95]. Common multivariate glyphs include iconographic displays such as star glyphs [53] or Chernoff faces [24, 55], and pixel oriented techniques [83] for mapping n-dimensional data sets to a visual representation.

In the star glyph [53], the data variable controls the length of the ray emanating from a central point. The rays are then joined together by a polyline drawn around the outside of the rays to form a closed polygon. Figure 4.11 shows a star glyph for each variable in Table 1.1 for the top three players. Note that each player has their own unique star representation, where the length of each axis in the star corresponds to a given measured variable.

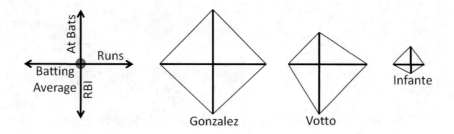

Figure 4.11: A star glyph for each of the top three players (by batting average) from Table 1.1.

Similar to the star glyph, the Chernoff face concept encodes data variables to form the shape of a face (as opposed to a star). The concept behind this representation is that humans are able to easily recognize faces and notice subtle changes with minimal difficulty. Facial features such as nose shape, ears, eyes and eyebrows are mapped to data variables. However, since features within faces vary in their perceived importance, it can be difficult to determine the appropriate mapping. Extensions to this concept include the work by Flury and Riedwyl [55]. Here, the authors suggested the use of asymmetrical Chernoff faces. Since faces are symmetrical about the vertical axis, only one half of the face would be needed to represent data. As such, the left half could be used to represent one set of variables, and the right half another. Studies on the effectiveness of the use of Chernoff faces have been conducted by Morris et al. [108], and results indicated that there were key features of the face that users studied the most including eye size and eyebrow slant. The study by Morris et al. also indicated that Chernoff faces may not have any significant advantage over other multivariate iconic visualization techniques.

Another method of visualizing multivariate data is the pixel-oriented technique described by Keim [83]. The idea of pixel-oriented visualizations is to represent as many objects as possible on the screen at one time by mapping each data variable to a pixel of the screen and then arrange the pixels in some intuitive manner. Key issues in pixel-oriented visualizations include the choice of mapping data values to color and the ordering of the data dimensions. In terms of color choices, Keim suggests the use of color over gray scales to increase the number of just noticeable differences that can be plotted on a screen. With respect to the ordering of pixels, if data sets have an inherent natural sequencing, then it is important that the visualization maintain that natural sequencing. When ordering the data dimensions, the problem is akin to that found in parallel coordinate plots where when one changes the order of dimensions, the resultant visualization will change as well. Keim proposes using similarity metrics to order the dimensions. Figure 4.12 shows a pixel oriented technique where each row and column of the data is colored based on an underlying derived metric. In this example, the growth rate across a sales category for 212 stores are being plotted. Stores with low growth rates are red, stores with positive growth rates are blue. Once the data is sorted, one can quickly begin finding groups of stores that are underperforming and then drill down into the data to search for other commonalities and anomalies.

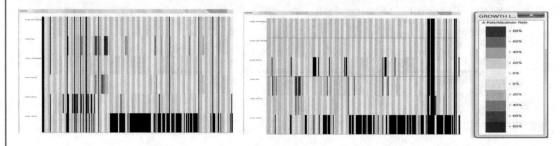

Figure 4.12: A pixel oriented technique showing the rate of sales over a one year period. The pixel oriented display on the left shows the data prior to sorting. Black values in the figure represent incomplete or missing data.

4.4   MULTIVARIATE ANALYSIS

While multivariate visual representations are capable of displaying data and helping users discover patterns, these displays are all still limited in the number of data dimensions that can be effectively shown. As such, the use of the visual representation alone is often not sufficient. Often times these techniques are coupled with dimensional reduction techniques such as principal component analysis, k-means clustering, multi-dimensional scaling, and self organizing maps in order to reduce the complexity of the data set. This allows the same visual techniques to be applied to a lower dimensional dataset, thereby reducing the clutter and more effectively showing the key properties of the data.

4.4.1   PRINCIPAL COMPONENT ANALYSIS

Principal Component Analysis [81] is one of the most commonly applied dimension reduction techniques. Principal component analysis is a deterministic analytical procedure that utilizes an orthogonal transformation to reduce a set of sample observations, with potentially correlated variables, into a set of uncorrelated variables called principal components. The number of principal components within a data set will always be less than or equal to the original number of variables in the sample set. The axis that maximizes the variance of all the projected points of the dataset is called the first component. The second component is orthogonal to the first and maximizes the variance of the projected dataset. The plane defined by the first two components will often provide the best overview of the dataset with respect to the data variance.

Given a data matrix of size $n \times m$, $\mathbf{X}$, each row of the matrix represents values corresponding to an observed sample, and each column represents a particular data variable. In order for principal component analysis to work properly, the mean of the dataset needs to be subtracted, thus $\mathbf{X}$ represents the zero mean data set. Next, the covariance matrix of the data set is calculated as

$$\mathbf{C} = (c_{i,j}, c_{i,j} = \text{cov}(\text{dim}_i, \text{dim}_j)) \tag{4.11}$$

The $\text{cov}(\text{dim}_i, \text{dim}_j)$ is defined as the covariance between two dimensions of the data set such that:

$$\text{cov}(X, Y) = \frac{\sum_{i=1}^{n}(X_i - \bar{X})(Y_i - \bar{Y})}{n - 1} \tag{4.12}$$

where $\bar{X}$ and $\bar{Y}$ are the means of their respective data dimensions.

The eigenvectors and eigenvalues are then calculated for the covariance matrix $\mathbf{C}$, and the eigenvectors calculated should be unit eigenvectors (i.e., vectors with a length of one). Once the eigenvectors are found, they are typically then ordered in terms of their eigenvalues. The eigenvector with the highest eigenvalue is the principal component of the data set. This vector shows the most significant relationship between the data dimensions. Note that the vector x is an eigenvector of the matrix $\mathbf{C}$ with an eigenvalue λ if the following equation is satisfied:

$$\mathbf{C}x = \lambda x \tag{4.13}$$

The eigenvectors corresponding to different eigenvalues are linearly independent, meaning that in an n-dimensional space, the linear transformation of $\mathbf{C}$ cannot have more the n eigenvalues. If $\mathbf{C}$ has less than n eigenvalues, then two or more of the variables within the space of $\mathbf{C}$ are correlated and the number of components needed to describe the dataset can be reduced.

Once the eigenvectors are calculated, a feature vector, $\mathbf{F}$ is then created. The feature vector is simply a matrix of all the eigenvectors. Note, that in creating the feature vector, the analyst may choose to remove eigenvectors that correspond to low eigenvalues as these vectors are thought to contain little information about the data. Once the eigenvectors are chosen and filtered, the original data set $\mathbf{X}$ is transformed such that:

$$\mathbf{X}' = \mathbf{F}^{\mathbf{T}}\mathbf{X}^{\mathbf{T}} \tag{4.14}$$

Thus, the original data is now transformed into a representation of its principal components.

The application of principal component analysis in creating visualizations is useful because it provides an analytical means of reducing data to its key components. In the multivariate visualization techniques shown before, there are a limited number of visual features that data can map to and still be perceptually differentiated. As such, it is critical to apply appropriate analyses to guide the visualization.

However, principal component analysis suffers from several pitfalls: it is very sensitive to outliers and to artifacts in the data [87]. Furthermore, methods that compute 2D projections of high-dimensional spaces are often difficult for users to understand. Analysts must now think about their data in the transformed space and attempt to apply their expert knowledge in this derived feature space.

4.4.2   K-MEANS CLUSTERING

Other data analysis techniques focus on finding similarities within the data sets and clustering the data based on these commonalities. K-means [66, 98] is one of the simplest unsupervised learning algorithms and is often applied as an initial solution for clustering. The k-means procedure classifies a given data set by using a user defined number of clusters, k, *a priori*. Initially, the k centroids are defined, one for each cluster. The centroids can be placed randomly, or algorithmically, but it should be noted that the initial placement will affect the result. The next step is to analyze each point within the data set and group it with the nearest centroid according to some distance metric. When all points have been assigned to a group, a new centroid for each group is calculated as a barycenter of the cluster, resulting from the previous step. Once the k new centroids are calculated, the algorithm reiterates through the data set, and each sample is again assigned to a cluster based on its distance to the new centroids. This process is repeated until the position of the centroids no longer change. Figure 4.13 shows the result of applying k-means clustering to the data set of Table 1.1.

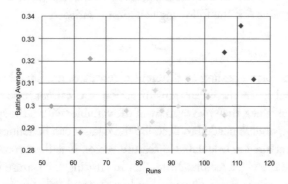

Figure 4.13: K-Means clustering of Runs versus Batting Average data from Table 1.1

More formally, the k-means algorithm attempts to minimize an objective function:

$$J = \sum_{j=1}^{k} \sum_{i=1}^{n} ||x_i^{(j)} - c_j||^2 \tag{4.15}$$

where $||x_i^{(j)} - c_j||^2$ is a chosen distance metric between the data sample $x_i^{(j)}$ and c_j.

While the k-means algorithm will partition the data into k clusters, there are several weaknesses. First, the clusters formed are very sensitive to the initial cluster centroid placement. A popular choice for seed placement is to simply generate a random seed location. However, the results produced also depend on the initial values for the means and often suboptimal partitions will be found. One common solution is to simply rerun the algorithm with a variety of initial seed locations. However, some seed locations can result in a cluster of size zero so that the cluster centroid can never be updated. This issue should be handled in any implementation of the code to avoid finding less than the k specified clusters.

Second, the results of the algorithm depend on the distance metric used. As the dimensionality of the data set grows, choosing an intuitive distance metric becomes difficult. Further, if the data set is sparse in certain dimensions, the resultant distance measures can negatively affect the clustering. One popular solution is to normalize each component of the distance metric by the standard deviation of the data within a dimension. One could also utilize a power transformation as a means of normalizing the data prior to performing analysis.

The final, and perhaps most critical, issue in applying k-means clustering is that the results depend on the value of k. Often times, there is no way of knowing how many clusters exist. However, recent work [43, 145] has shown that the relaxed solution of k-means clustering is given by the principal component analysis. Thus, the subspace of the data spanned by the principal component directions is identical to the cluster centroid subspace.

As in principal component analysis, the resulting analysis for the k-means clustering can be used to enhance the visualization. However, the clusters found in n-dimensional space still need to be projected into a lower dimensional space in order to visualize them. As in any projection, information will be lost and clusters will overlap. Some work has focused on clustering over a subset of dimensions and then projecting these clusters into the space of variables that were not used in clustering. Hargrove and Hoffman [67] took ecological data and clustered the data by their ecological measures, ignoring the spatial locations. The clusters in this ecological space were then projected down to the geographical space, and the analyst could then visualize where certain ecotopes formed with respect to the geography. Figure 4.14 illustrates this clustering and projection concept on a three-dimensional dataset consisting of patient health records in the state of Indiana. In this figure, one can see that certain groups of syndromes map to various areas in the state. These areas could be further analyzed for commonalities and other data sources could be utilized as a means of correlating results across various spatial locations.

Recent work has focused on the use of interactive techniques as a means of aiding users to better understand dimensional reduction techniques. Work by Jeong et al. [79] and Choo et al. [28] both explore the addition of combined visualizations and interactivity as a means of helping users understand the complex combination of dimensions done in dimensional reduction. The addition of interactivity has been shown to be effective in helping users understand the relationship between the data and the calculated dimensional reduction.

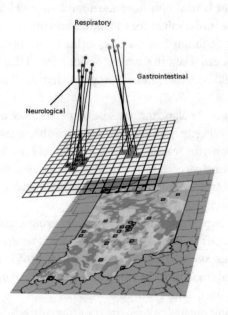

Figure 4.14: K-means clustering in one subspace projected to another subspace. In this figure, the clustering is done with respect to patient syndromes, and the projection is done to the syndrome locations.

4.4.3 MULTI-DIMENSIONAL SCALING

Another means of projecting high-dimensional data into lower-dimensional representations is through the application of a multi-dimensional scaling method [89, 101, 117]. In the classical, multi-dimensional scaling scheme, the pairwise distances between observations is preserved in lower-dimensional projections. In this manner, analysts can learn about differences in the high-dimensional space by viewing their relative distances in a two-dimensional projection.

Given two observations (x_i, x_j), the multi-dimensional scaling procedure finds points in a lower dimensional projection (z_i, z_j) such that the error of the distance between two points in the projected space is minimized. These distances, $\delta_{i,j}$ are the entries in the dissimilarity matrix, which is defined as:

$$\Delta_{m,n} = \begin{pmatrix} \delta_{1,1} & \delta_{1,2} & \cdots & \delta_{1,n} \\ \delta_{2,1} & \delta_{2,2} & \cdots & \delta_{2,n} \\ \vdots & \vdots & \ddots & \vdots \\ \delta_{m,1} & \delta_{m,2} & \cdots & \delta_{m,n} \end{pmatrix} \qquad (4.16)$$

The goal is to then find a set of vectors, z, such that

$$||z_i - z_j|| \approx \delta_{i,j} \tag{4.17}$$

Equation 4.17 is typically referred to as a stress function, and multi-dimensional scaling algorithms attempt to minimize the stress function.

In interpreting a multi-dimensional scaling projection, it is important to realize that the axes are meaningless. The axes are created such that the points chosen in this space represent the relationship between points in the higher dimensional space. Furthermore, the orientation of the space is arbitrary. Finally, the analyst must also keep in mind that the distances between the objects in this space are distorted representations of the relationships within the data. Clusters in the projection are groups of items that are closer to each other in the multi-dimensional space than they are to other items. When tight, highly separated clusters occur; it may suggest that each cluster is a region of the data that should be analyzed individually.

Other variations of multi-dimensional scaling exist, including weighted multi-dimensional scaling. Here, the dimensions of the high-dimensional space are weighted in order to express their relevance in the transformed space. The weights provide the analyst with a means to incorporate expert knowledge about the relationships between data and can ultimately be used to guide the underlying visualization.

4.4.4   SELF-ORGANIZING MAPS

Another means of producing a low-dimensional representation of high-dimensional data is the self-organizing map introduced by Kohonen [86]. Self-organizing maps are a type of artificial neural network that utilize unsupervised learning to reduce the data dimensionality. First, a network is created from a 2D lattice of nodes, each of which is directly connected to the input layer. Figure 4.15 shows a 3×3 network of nodes connected to an input layer. Each node within the network will have its own unique coordinate position (x, y) and a vector of weights, W. The dimensionality of this vector is equal to that of the high-dimensional input.

The node weights are created from an initial distribution of random weights, and the self-organizing map algorithm iteratively modifies the weight until a stable solution is reached. Next, each sample in the input X_i is iterated through, and a distance metric between each sample point and each node's weight is calculated, and the node with a weight vector that is closest to the input vector is found. This node is then tagged as the best match for the given input.

Once all the input vectors have been assigned to a node, the next step is to determine which nodes should be neighbors. In the initial step, the radius of the neighborhood is predefined and a typical choice is to make the radius equal to half the network size. For example, if the network size is 4×4, then the initial radius, σ_0 will be 2. As the algorithm progresses over time, the radius of the neighborhood will decrease according to an exponential decay function:

$$\sigma(t) = \sigma_0 \exp(-\frac{t}{\lambda}) \tag{4.18}$$

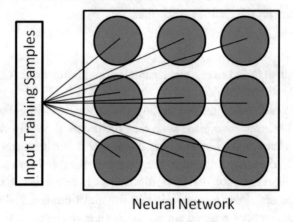

Figure 4.15: Self-Organizing map network.

λ is the time constant for decay and is based on the number of maximum iterations that the self-organizing map algorithm will run. Thus, for each node in the lattice, the radius of its neighborhood is known for each iteration.

Given the location of all other nodes in the lattice and the radius, for each node, the algorithm iterates through all other nodes to determine if these nodes are within its neighborhood. If a node is found to be within the neighborhood, then the weight vector of the node is adjusted such that

$$W(t+1) = W(t) + \Theta(t)L(t)(X(t) - W(t))$$ (4.19)

In this equation, t represents the time step, $L(t)$ represents the learning rate which decreases with time, and $\Theta(t)$ scales the learning rate based on each node's distance. The learning rate is defined as:

$$L(t) = L_0 \exp(-\frac{t}{\lambda})$$ (4.20)

The learning rate scale factor is defined as:

$$\Theta(t) = \exp(-\frac{\text{dist}^2}{2\sigma^2(t)})$$ (4.21)

As with principal component analysis, k-means and multi-dimensional scaling, self-organizing maps are commonly used means of data dimensionality reduction for visualization. The underlying analysis can make it easy for an analyst to see relationships between the underlying high-dimensional data components. Self-organizing maps may be thought of as a nonlinear generalization of principal component analysis. Issues in using self-organizing maps arise due to the fact that each generation of the map, using randomized initial seed weights, has the potential to generate a new result. Self-organizing maps are also highly sensitive to missing data, and they are computationally expensive.

4.5   TIME SERIES VISUALIZATION

The previous techniques and analysis focused on dimensional reduction and displaying the relationship between multivariate data within some lower projection. In those examples, one of the most common multivariate data structures was ignored, time. The analysis of time series data is one of the most common problems in any data domain, and the most common techniques of visualizing time series data (sequence charts, point charts, bar charts, line graphs, and circle graphs) have existed for hundreds of years. Furthermore, widespread interest in discovering time series trends has created a need for novel visualization methods and tools.

The goal of time series visualization is to put into context the relationship of the past, present and (potentially) the future. In order to understand what is currently happening now, it is often helpful to try to understand the historical concepts. Often, daily questions relate current phenomenon to what has happened in the past. "Is this year hotter than the last?" "Has crime increased in my neighborhood?" These questions are predicated on knowing the historical context of data.

4.5.1   LINE GRAPHS

Perhaps the most common visual representation used to answer such questions and display the changes of events over time is the line-graph. One can imagine the line graph as a specialized form of the histogram where each point on the line represents a binning of some events over an increment of time. For example, in Figure 4.16, a visualization of the number of coastguard search and rescue cases in the Great Lakes from January 1, 2004 to December 31, 2004 is shown. Line graphs are able to provide the analyst with a quick view of the fluctuations over time allowing them to examine trends.

Depending on the aggregation within a line graph, noise can be smoothed, allowing seasonal trends to emerge. Figure 4.16 illustrates the effects of aggregating data by day and by week. In Figure 4.16(Left), the data is binned by day and plotted over time. Here, the noise in the data tends to obscure some of the weekly variations. By binning the data by a week (as opposed to by day), the seasonal summer pattern is highlighted and the noisy variations are obscured.

There are a number of variations on the standard line graph, and depending on the questions the analyst wishes to ask of their data, different representations may be better suited. For example, the theme river visualization introduced by Havre et al. [69] attempts to show thematic changes within a time stream. This work creates colored "currents", building upon the river metaphor. Here, the visualization assumes that the user is less interested in the individual documents and more interested in how the themes change over the whole collection. Thus, given a set of documents, the occurrence of given keywords over time will be plotted as a stream. Each stream is given its own unique color and the thickness of the stream varies based on the occurrence metric calculated.

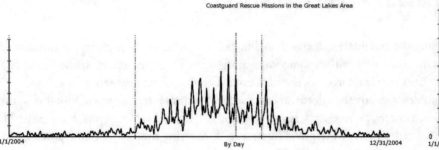
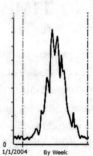

Figure 4.16: Time series visualization showing the number of Great Lakes search and rescue missions occurring over 2004 binned by day and by week.

4.5.2   CYCLICAL TIME

While understanding themes and seasonal trends is important, it can be difficult to compare patterns between years in a line graph visualization. Typically, one thinks of time as linear, and this corresponds to our natural perception of time; however, hours of the day, days of the week and seasons of the year are cyclical in nature. Figure 4.17 shows the potential implications that changing the order of cyclical time components can have. In Figure 4.17 (Left), the average patients seen in an Emergency Department for each day of the week (Monday - Sunday) is displayed as a histogram. In Figure 4.17 (Right), the average patients seen in an Emergency Department for each day of the week are shown again; however, this time they are ordered from Friday - Thursday, allowing an analyst to better visualize that peak visits to this Emergency Department occur on Sunday and Monday.

Thus, depending on the question the user wishes to ask of the data and the manner in which the data is to be interpreted, different visual representations of the time series should be created. One recent technique for exploring cyclical time includes the Time Spiral[72, 139]. In this method, each aggregate time unit maps to a curved arc on a spiral, and the spiral is designed such that every cycle is the length of one loop of the spiral. By doing this, spiral segments align allowing users to explore cyclical temporal correlations.

4.5.3   CALENDAR VIEW

Another method that attempts to show cyclical patterns is the calendar view visualization developed by van Wijk and Selow [137]. In the calendar view (Figure 4.18), entries in the calendar bin the data by day, and then the columns and rows are binned to illustrate distributions by day of the week and a chosen temporal cycle. This visualization provides a means of viewing data over time, where each date is shaded based on the overall yearly trends. Here, the max data value is shaded the darkest blue, and the lowest data values are shaded white. Users can interactively control the cycle length of the calendar view. Figure 4.18 gives a calendar view visualization of the number of noise complaints

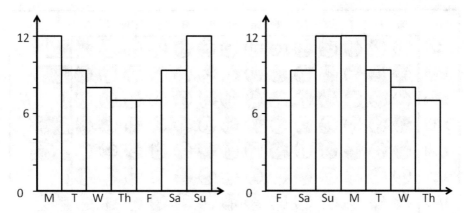

Figure 4.17: Histograms of the average number of patient visits to an emergency department by day of the week. The left image starts the weekly cycle on a Monday, where the right image starts the cycle on a Friday.

in the West Lafayette Police District in the year 2000. A histogram appears on the bottom and right-hand side of the calendar view, providing users with a quick overview of both daily patterns as well as overall weekly trends. The coloring of the calendar view is done to highlight daily peaks, where darker colors indicate a larger number of complaints. Analysts can quickly look for patterns across time by day and week using these sorts of tools.

While these types of visualizations allow one to compare trends of data across time, analysts often have other questions of the data that are concerned with how the correlation between variables has changed over time as well. If one wants to examine the relationship between family income and health reports from census data, a scatterplot visualization would allow the analyst to explore the correlation of these two variables. However, if one wishes to explore the change of these variables over time other visual representations and interaction techniques need to be explored.

4.5.4   MULTIVARIATE TEMPORAL EXPLORATION

One common means of exploring a variety of variables across time is the use of small multiples [134]. Small multiples are a series of smaller pictures that detail differences across categories. These pictures are of consistent design and should be arranged in a meaningful structure that encourages comparison. For time series visualization, small multiples can consist of small line graphs plotted for each variable. The concept of small multiples is akin to the Trellis display [14]. The Trellis display was a framework created for displaying large multivariate data, and it is often useful in initial analysis stages when data models are formulated or during the diagnostics stage when models are evaluated. A Trellis display consists of panels laid out into a three-way rectangular array consisting of columns, rows and pages of data. Each panel of the Trellis display (or small multiple) visualizes one data variable with respect

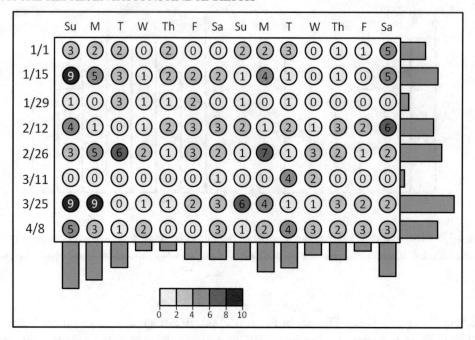

Figure 4.18: A calendar view showing the daily number of noise complaints in the West Lafayette Police District in the year 2000. Bins on the bottom and side of the calendar view provide quick look histograms of the summation of the calendar columns and rows. The coloring in the calendar view is used to visually expose days with high and low data counts in order to allow the user to discern potential patterns and better understand their data.

to a particular condition. In the case of temporal data, each panel in the Trellis display would show a variable of interest plotted over time.

Another visualization method that looks at representing multiple variables with respect to time is the time wheel [131]. This method is akin to parallel coordinate plots in that each variable is represented as a plane. These planes are radially arranged around an exposed central time axis. At each time point, a line to the value of the variable at time t is drawn.

4.5.5 ANIMATION

While both small multiples and the time wheel allow for multivariate comparison over time, both methods require a large amount of comparisons which may increase the cognitive burden. However, given that time moves in a linear fashion, one can animate data to show the movement of trends over time. Griffin et al. [61] compared the use of animation in maps to the use of small multiples displays. Their work found significant advantages in both the speed and accuracy of identifying clusters and trends in the data.

In fact, many questions asked about temporal data have to do with the change throughout time. These patterns are formed by the combination of four characteristics [54]: the magnitude of the change, the shape of the change, the velocity of the change, and the direction of the change. These patterns can often be explored through the use of animation. Some means for tracking animation and enhancing scatterplots include visualizing the 'trails' that the data leaves on the plots over time, as well as encoding the size of the points in the graph as a third variable dimension.

Recent work in temporal animation includes animated scatterplots. These plots display the correlation between two variables, and by animating the relationship over time. Recent work by Robertson et al. [116] evaluated the effectiveness of animation for trend visualization. In this work, the authors utilized both animated scatterplots and small multiples. Results from their study indicated that for analysis, static plots proved more effective; however, when communicating results, animation can be a more effective means of conveying information.

As such, it is important to determine what questions an analyst needs to ask of their temporal data prior to creating the visualization. Both animation and interactive techniques can aid the analyst in exploring patterns and drilling down into the data. However, if the analyst can provide definitions of anomalies that they are searching for within the data, analytic techniques can be applied to enhance the underlying visual representation.

4.6   TEMPORAL MODELING AND ANOMALY DETECTION

Time series analysis are methods that model and extract meaningful characteristics and statistics from temporal data. What makes time series analysis different then other data analysis problems is the fact that each sample has an inherent ordering. Furthermore, one can use the fact that observations close together in time will generally be more closely related than those farther apart.

4.6.1   CONTROL CHARTS

One of the most common methods of identifying anomalies within temporal data is through the use of control charts. Control charts consist of data points representing a given statistic from a process at different times within the data. Typically, the mean and standard deviation of the data is calculated for some predefined historical window. If the current value is greater than some pre-set number of standard deviations from the mean, then an alert is generated (Shewart suggested three standard deviations [120], which is the basis for the Shewart control chart).

Here, one may wonder why visual analytics would be necessary if alerts can be determined automatically. If we explore finding temporal alerts for healthcare data, the need for visual analytics becomes more apparent. In the state of Indiana, approximately 78 emergency departments report 9 syndromic signals to the state department of health. Given the inherent noise of the data, the resultant 702 time series generate on the order of 100 alerts per day. The majority of these alerts are false positives that require investigative effort to confirm or deny after the statistical anomaly is detected. By incorporating enhanced visual and multiple analytical algorithms, systems can be created to help reduce the number of alerts and reduce the amount of time need to explore each alert.

The purpose of control charts is simply to allow an analyst to detect changes of events that are indicative of an actual process change. If the analyst has knowledge of the process, then their expert guidance can help rule out anomalies based on factors that are extrinsic to the control chart process. Typically, control charts are built with varying limits in order to provide analysts with early notifications if something is awry. By adding in such warning limits, an analyst can determine whether special cases have affected the process or if an early stage investigation needs to take place.

As with earlier analytical methods, the control chart methods often perform best when the underlying data is normally distributed. Thus, the application of an appropriate power transformation can potentially improve the underlying analysis. Here, it is important to realize that different control charts provide different levels of detail to the analyst. For example, Shewart control charts are good at detecting large changes, but not good at small changes. For small changes, other types of control charts are preferable (for example, exponentially weighted moving averages and cumulative summation [105].

The EWMA algorithm takes the temporal data and first smoothes the time series:

$$S_t = (1 - \alpha)S_{t-1} + \alpha y_t, \ \text{for } t \geq 2,$$

where S_t is the smoothed series, y is the original time series and $0 \leq \alpha \leq 1$. To initialize the smoothed series at $t = 1$, we use $S_0 = \bar{y}$; that is, the sample mean:

$$\bar{y} = \frac{1}{n} \sum_{t=1}^{n} y_t$$

Given the smoothed series, we calculate an upper control limit (UCL) as:

$$UCL = \bar{y} + L\sigma \sqrt{\frac{\alpha}{2 - \alpha}[1 - (1 - \alpha)^{2i}]}$$

where σ is

$$\sigma = \sqrt{\frac{\frac{\sum_{t=2}^{n} |y_t - y_{t-1}|}{n - 1}}{1.128}}$$

where 1.128 is a table value from the original EWMA design [105].

Requirements in initializing this algorithm include an appropriate choice of α. This parameter weights each previous sample and recommended values range from $.05 \leq .3\alpha$ [105]. The second key parameter is that of L. This value establishes the control limits and is typically set to 3 to match the Shewart chart limits.

As previously stated, the exponentially weighted moving average chart is sensitive to small shifts in the process; however, it fails to match the ability of the Shewart chart to capture larger shifts. Thus, one could consider combining both processes as a means of corroborating signals to detect both large shifts and shorter anomalies.

Another method that has been shown to be more efficient at detecting small shifts is the cumulative summation control chart.

$$S_t = max\left(0, S_{t-1} + \frac{X_t - (\mu_0 + k\sigma_{x_t})}{\sigma_{x_t}}\right) \tag{4.22}$$

Equation 4.22 describes the CUSUM algorithm, where S_t is the current CUSUM, S_{t-1} is the previous CUSUM, X_t is the count at the current time, μ_0 is the expected value, σ_{x_t} is the standard deviation, and k is the detectable shift from the mean (i.e. the number of standard deviations the data can be from the expected value before an alert is triggered). Often times, a sliding window will be employed to calculate the mean, μ_0 , and standard deviation, σ_{x_t}, with a temporal lag, meaning that the mean and standard deviation are calculated over the previous q time samples n samples prior to the current sample under analysis. Such a lag is used to increase sensitivity to continuous anomalies while minimizing long term historical effects.

4.6.2   TIME SERIES MODELING

While control chart methods enable anomaly detection, it is often the case that analysts wish to predict future time series trends. The most common methods of time series modeling include the autoregressive moving average (ARMA) and the autoregressive integrated moving average (ARIMA) methodology developed by Box and Jenkins [17]. These models are fitted to time series data as a means of forecasting future data points. Typically, the ARMA model is defined with respect to variables p and q, while the ARIMA model is defined with respect to variables p, d, and q, which are non-negative integers that refer to the order of the autoregressive (p), integrated (d) and moving average (q) portions of the model.

Thus, given a time series, X_t, the ARMA model is defined as:

$$\left(1 - \sum_{i=1}^{p} \alpha_i L^i\right) X_t = \left(1 + \sum_{i=1}^{q} \Theta_i L^i\right) \epsilon_t \tag{4.23}$$

Here, L is the lag operator meaning that

$$L^i X_t = X_{t-i}. \tag{4.24}$$

α_i are the parameters for the autoregressive portion of the model, Θ_i are the parameters for the moving average terms, and the ϵ_t is the error term. Typically, the error term should be independent, identically distributed Gaussian white noise. Similarly, given a time series, X_t, the ARIMA model is defined as:

$$\left(1 - \sum_{i=1}^{p} \phi_i L^i\right)(1 - L)^d X_t = \left(1 + \sum_{i=1}^{q} \Theta_i L^i\right) \epsilon_t \tag{4.25}$$

In ARIMA and ARMA models, a parametric approach is utilized for time series forecasting. As an alternative, one may also employ non-parametric modeling methods. One such method is seasonal-trend decomposition based on *loess* (locally weighted regression) [34]. Seasonal-trend decomposition is used to separate the time series into its various components. Seasonal trend decomposition components of variation arise from smoothing the data using moving weighted-least-squares polynomial fitting, in particular *loess* [33], with a moving window bandwidth. The degree of the polynomial is 0 (locally constant), 1 (locally linear), or 2 (locally quadratic).

Here, it is important to note that in order to appropriately model the time series using STL, the mean and variance of the data needs to be independent. To accomplish this, a power transformation can be applied to the data as previously described in Chapter 3. In cases where the data consists of counts following a Poisson distribution, a square root transformation will make the mean independent of the standard deviation.

For a given time series consisting of daily samples, one can decompose the counts into a day-of-the-week component, a yearly-seasonal component that models seasonal fluctuations, and an inter-annual component which models long term effects:

$$\sqrt{Y_t} = T_t + S_t + D_t + r_t \tag{4.26}$$

where for the t-th day, Y_t is the original series, T_t is the inter-annual component, S_t is the yearly-seasonal component, D_t is the day-of-the-week effect, and r_t is the remainder.

The procedure begins by extracting the day-of-the-week component, D_t. First, a low-middle frequency component is fitted using locally linear fitting. Then D_t is the result of means for each day-of-the-week of the $\sqrt{Y_t}$ minus the low-middle-frequency component. Next, the current D_t is subtracted from the $\sqrt{Y_t}$, and the low-middle-frequency component is re-computed. This iterative process is continued until convergence. After removing the day-of-the-week component from the data, loess smoothing is applied to extract the inter-annual component, T_t, using local linear smoothing. Finally, loess smoothing is applied to the data with the day-of-week and inter-annual components removed, thereby obtaining the yearly-seasonal component, S_t, using local quadratic smoothing. After removing the day-of-week, inter-annual, and yearly-seasonal components from the time series, the remainder is found and should be adequately modeled as independent identically distributed Gaussian white noise, indicating that all predictable sources of variation have been captured in the model. Figure 4.19 illustrates the resultant signal components extracted when applying seasonal trend decomposition to daily patient counts in an Emergency Department.

Once the data has been modeled, one can utilize the model for future prediction. For prediction using the STL method, the analyst may rely on some statistical properties of loess, namely that the fitted values $\hat{Y} = (\hat{Y}_1, \ldots, \hat{Y}_n)$ are a linear transformation of the observed data, $Y = (Y_1, \ldots, Y_n)$. Each step of the STL decomposition involves a linear filter of the data. In other words, an output time series $x = \{x_1, \ldots x_n\}$ is produced by an input time series $w = w_1, \ldots, w_n$ through a linear

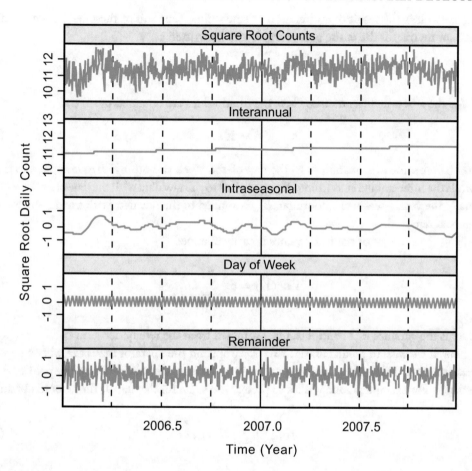

Figure 4.19: The extracted temporal components of and Emergency Department time series using seasonal trend decomposition.

combination

$$x_i = \sum_{i=1}^{n} h_{ij} w_j. \tag{4.27}$$

If we let H be a matrix whose (i, j)-th element is h_{ij}, then we have

$$x = Hw. \tag{4.28}$$

H is now referred to as the operator matrix of the filter. Now, let H_D, H_S, and H_T denote the operator matrices of the day-of-week, yearly-seasonal, and inter-annual filters, respectively. All of these matrices are $n \times n$. H_S and H_T are straightforward to calculate [34], but H_D is more difficult to calculate as it is the result of an iteration of smoothing. Once all of these have been calculated, the operator matrix for the entire procedure, H can be written as

$$H = H_D + H_T(I - H_D) + H_S(I - H_D - H_T(I - H_D)), \tag{4.29}$$

where I is the $n \times n$ identity matrix. Now, the fitted values are obtained by

$$\hat{Y} = HY. \tag{4.30}$$

To make better sense of Equation 4.29, the day-of-the-week smoothing, H_D, is applied to the raw data, while the inter-annual smoothing, H_T, is applied to the raw data with the day-of-week removed, and finally, the yearly-seasonal smoothing, H_S, is applied to the raw data with the day-of-week and inter-annual removed.

Now, the variance of the fitted values is easily obtained

$$Var(\hat{Y}_i) = \hat{\sigma}^2 \sum_{j=1}^{n} H_{ij}^2, \tag{4.31}$$

where $\hat{\sigma}^2$ is the variance of Y, and it can be estimated from the remainder term r_t.

Now, if we wish to predict ahead, x days, we append the operator matrix H with x new rows, obtained from predicting ahead within each linear filter and use this to obtain the predicted value and variance. For example, if one wishes to predict the value for day $n + 1$, they would obtain

$$\hat{Y}_{n+1} = \sum_{j=1}^{n} H_{n+1,j} Y_j \tag{4.32}$$

and

$$Var(\hat{Y}_{n+1}) = \hat{\sigma}^2 (1 + \sum_{j=1}^{n} H_{n+1,j}^2), \tag{4.33}$$

so that an approximately 95% prediction interval will be calculated as

$$\hat{Y}_{n+1} \pm 1.96 \sqrt{Var(\hat{Y}_{n+1})}. \tag{4.34}$$

It should be noted that the techniques described in this section are complex modeling methods. Poor parameter choices and bandwidth selections can critically alter the resultant predictions. As such, care should be taken when employing such models, and advanced descriptions and details are left to texts devoted specifically to the subject. However, while the methods are complicated, the means of predicting data can greatly enhance the resultant analytical process, potentially providing a means from moving from visual analytics to predictive visual analytics.

4.7   GEOGRAPHIC VISUALIZATION

Amongst the previously explored visualization and analysis methods, there has been an underlying concept of space. Temporal events occur at a place, and variables are mapped to a location. In fact, one can see that questions about time often are entwined with location, for example, "Has crime increased in my neighborhood?" Thus, a key component that is found in many multivariate data sets is a geospatial component.

Geographic visualization utilizes sophisticated, interactive maps to explore information, guiding users through their data and providing context and information in with which to generate and explore hypotheses. In more recent years, it has ballooned to include increasingly complex data, other spatial contexts, and information with a temporal component. Relevant summaries on work in the field can be found in texts by MacEachren [96], Peuquet [112], Dykes and MacEachren [46], and Andrienko and Andrienko [7]. These books detail thoughts on knowledge discovery and the exploratory analysis of spatial and temporal data. Other reviews on spatiotemporal data mining and analysis can be found in [4, 5, 47].

Many of the geovisualization practice methods described in those textbooks [7, 46, 112] have been leveraged to create systems for data exploration, combining both interactive mapping techniques and statistical methods. Dykes et al. [45] utilized web services to deliver interactive graphics to users and introduced an approach to dynamic cartography supporting brushing and dynamic comparisons between data views in a geographical visualization system. Andrienko et al. [6] developed the Descartes system in which users were able to select suitable data presentation methods in an automated map construction environment. Work by Hargrove and Hoffmann [67] used multivariate clustering to characterize ecoregion borders. Here, the authors select environmental conditions in a map's individual raster cells as coordinates that specify the cell's position in environmental data space. The number of dimensions in data space equals the number of environmental characteristics. Cells with similar environmental characteristics will appear near each other in dataspace. Edsall et al. [49] created a system that decomposed geographic time series data using a three-dimensional Fourier transformation of the geographic time series, allowing users to explore the three-dimensional representation of the physical and spectral spaces. Carr et al. [22] utilized a two-way layout of choropleth maps to enhance a user's ability to compare dataset and explore data hypotheses.

Figure 4.20 shows several common methods of geographical visualization. For a simple overview of the spatial distribution of events, one can plot all the locations on a map (Figure 4.20 - Left). However, as the data set becomes large, the points will become cluttered and details will be obscured. Another method is to simply aggregate the data over a geographical region and then color code each region to the appropriate aggregate level (Figure 4.20 - Middle). Again, such a method can remove details and often times, the geographical boundaries for aggregate regions are do not map to geographically meaningful structures. Finally, one can also consider applying density estimation to the points to visualize the sample distribution (Figure 4.20 - Right). However, appropriate choice of bandwidth selection and the fact that density estimation will ignore natural boundaries (such as rivers) can obfuscate the analysis.

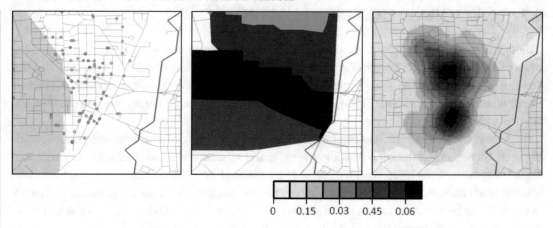

Figure 4.20: Sample geographic visualizations of noise complaints in West Lafayette, Indiana. (Left) Plotting events as glyphs on the map. (Middle) A choropleth map of noise distribution over census tracts. (Right) A kernel density estimation of the noise distribution.

4.7.1   CHOROPLETH MAPS

Of the examples shown in Figure 4.20, one of the most commonly employed geographical visualization technique is the choropleth map. Choropleth maps [143] are maps in which areas of the map are shaded in proportion to a measured variable. Coloring is based on a classification (histogram binning) of the distribution of the measured variable. The number of colors depends on the number of classes (bins). Too many classes can overwhelm the user and distract them from seeing trends; furthermore, an overabundance of classes can compromise the legibility as colors become difficult to distinguish. Thus, one can see that the visual representation of the choropleth map is highly influenced by the class interval selection.

Along with the complexity of choosing categories for choropleth map colors, an another key issue in creating maps is the fact that not all areas of the map are created equal. Size is a dominant visual cue, and manmade geographical boundaries vary wildly in their overall area (compare the size of Texas to Rhode Island, for example). Thus, giving equal representation to all areas is an inherent problem with choropleth maps when the geographic areas vary greatly in size. These different sizes can hide changes in the data or draw attention to unimportant areas of the map. Furthermore, when aggregating data, small areas (like major cities) may overwhelm the data of larger regions (like states). Such aggregation problems also give rise to ecological fallacies as analysts may be prone to making inferences on their data based on the aggregate region.

4.7.2   DASYMETRIC MAPS

In an attempt to rectify some of the issues of aggregating data, cartographers have proposed the use of dasymetric maps. A dasymetric map depicts statistical data, but it uses boundaries that divide the mapped area into zones of relative homogeneity. The purpose of such divisions is to attempt to better display the underlying statistical surface.

Dasymetric maps are akin to choropleth maps in that the variable being displayed is broken into a set of classes by color. However, the boundaries in the dasymetric map are based on sharp changes in the statistical surfaces being mapped, while boundaries within choropleth maps are based on units established for more general (and typically governmental) purposes. The zones developed in the dasymetric map are developed to be homogenous, while choropleth zones are not defined based on the data and have varying levels of internal homogeneity [50]. In order to create a dasymetric map, ancillary data files are needed for the interpolation of data from the original zones (state, county, zip code, etc.) of the choropleth maps. Often times, this data is in the form grid and polygonal data that are broken into homogenous areas.

4.7.3   ISOPLETH MAPS

While choropleth and dasymetric maps attempt to aggregate data into various classes, isopleth maps attempt to fit data to a continuous distribution. Such maps are typically used for mapping things such as surface elevation, precipitation and other such measures. Once the distribution is calculated, the measurements are typically shown by an overlay of lines (called isopleths), which connect points of equal value. Isopleths will never cross each other as that would mean that the measurement at that location would have two different values. Isopleths are equally spaced statistically; thus, in areas where isopleths are drawn near to each other, there is a rapid change in value. There are several mathematical methods for calculating the continuous distribution needed for isomap creations. Density estimation (as discussed in Section 4.2) is one such common method. In density estimation, the distribution is estimated based on the distance from sample points (where the distance is based on the bandwidth). Another approach that can be used is *kriging* [38].

Ordinary kriging is a linear least squares estimation technique that attempts to estimate the value of an unknown real-valued function from a set of measured samples. Instead of weighting data points by their distance, ordinary kriging utilizes the spatial correlation within the data to determine the weighting. Thus, given a measurement $f(x_i)$ at location x_i (where x_i is a latitude, longitude pair in geospace), an estimate at unknown locations can be calculated as

$$\hat{f}(x^*) = \sum_{i=1}^{n} \lambda_i(x^*) f(x_i) \tag{4.35}$$

Here, the λ_i are the solutions to a system of linear equations which are obtained by assuming that f is a sample-path of a random process $F(x)$ where the error of this prediction is given as:

$$\epsilon(x) = \hat{f}(x) - \sum_{i=1}^{n} \lambda_i(x) \hat{f}(x_i) \tag{4.36}$$

One advantage of kriging is that the estimated values have a minimum error associated with them, and this error can be quantified (and even visualized).

More advanced methods of kriging can also be employed. Techniques such as indicator kriging take a non-parametric approach to estimating distributions. Like ordinary kriging, the correlation between data points determines the values in the model distribution; however, indicator kriging makes no assumption of normality. As with other techniques described in this chapter, an appropriate use of the power transformation could be used to bring data in line with model assumptions and improve the output.

4.7.4   CLASS INTERVAL SELECTION

In each of the mapping techniques, choosing the range of data value to represent as colors or distances between isopleths is perhaps the most challenging design choice one will make. The problem of class interval selection is equivalent to the problems faced in histogram binning. Monmonier [103] states that poorly chosen intervals may convey distorted or inaccurate impressions of data distribution and do not capture the essence of the quantitative spatial distribution. As such several simple class interval selection/binning methods (such as quantile, equal interval and standard deviation) and more complex methods (natural breaks [78], minimum boundary error [39] and genetic binning scheme [11]) have been used traditionally [103]. Several researchers have reported the comparative utility of these methods. Smith [124] reported that quantile and standard deviation methods were most effective with normally distributed data and were most inaccurate with asymmetrical and/or peaked distributions. Moreover, equal interval and natural breaks methods were inconsistent for various data distributions. Frequency based plots have been used to delineate class intervals [100], particularly for datasets with a standard normal distribution with the curve split into equal bins based on mean and standard deviation [10]. As with histogram binning, the output from class selection methods is greatly influenced by the underlying data distribution.

However, as we saw with histogram visualizations, most visual analytic tools rely either on a default bin selection or a user controlled selection. However, the default selection can lead to poor choices in bin width selection, and users need to be familiar with the underlying data distribution in order to obtain an effective colormap. Therefore, an automatic classification method could be favorable when data distributions change frequently, as is the case in interactive visual analysis environments. An automatic color binning/classification method based on extracted statistical properties, including skewness, was described by Schulze-wollgast et al. [118]. However, they limited the choice of classification to just the logarithmic and exponential mappings, which may not be the best choice for every dataset. As such, the power transformation, with an appropriate power value that is best able to reduce skewness and condition the data to near normality, could be beneficial in interactive environments to provide an automatic initial visualization based on the transformed data.

Furthermore, in the case of skewed data, research has shown that traditional methods, such as equal interval classification, is ineffective at aiding users in identifying clusters and rates of change in choropleth maps due to inaccurate binning [124]. However, research showed [19, 124] that equal interval classification is as effective as the more sophisticated binning/classification schemes (e.g.,

Jenks natural breaks [78] and minimum boundary error [39]) when the data falls under a normal distribution.

Given the fact that interval classification for choropleth maps can lead to quantization errors, there has been much debate as to whether or not choropleth maps should use classes at all. Detailed reviews of this debate can be found in Brewer and Pickle [19] and Slocum [123]. One side of the argument states that continuous color scales should be used as a continuous color scale will not lead to data quantization errors that are inherent when choosing class intervals. The other side argues that users are not able to discern all the values in a continuous color scale and such mappings will reduce the time and accuracy in reading a map. Further details on designing better maps can be found in work by Dykes and Unwin [48] and Brewer [18].

4.7.5   INTERACTIVE MAPS

One means of overcoming issues, in class interval selection, is through the use of an interactive visual environment. By employing interaction techniques such as brushing, linking and drill-down operations, users can now explore areas on the map and retrieve the exact data values. Harrower [64] notes that the use of interactive techniques for retrieving data from maps has several advantages over traditional static maps. The first is that such interactivity will allow for greater precision. When analyzing a static map, colors represent a range of value, for example, one might know that the population in a region lies between 40,000 and 50,000; however, the addition of interactivity would allow one to find the exact value of the region is 42,317. Second, this interaction adds not only precision, but also provides users with a quicker means of data retrieval. Attention is no longer split between looking at the map and looking at the legend; instead, users can click regions to retrieve data values. As such, interactive maps have become a staple of many interactive visual analytics systems.

4.7.6   ANIMATING MAPS

The previously described visualization techniques focus on the modeling and projection of statistical measures over a geographical location. However, as both spatial location and temporal variations are critical components in answering analytical questions. One means of exploring spatiotemporal variations, using statistical mapping techniques, is the addition of animation. Work in this area focuses on the animation of choropleth maps and the ability of a user to explore and extract patterns from a given animation. Figure 4.21 shows three time steps of an animated choropleth map, with the color representing the number of individuals infected by a simulated pandemic. By animating such models and simulations, the rate of flow and fluctuations within the model can be easily understood.

Yet, the animation of choropleth maps brings with it a series of new challenges. In previous sections, a discussion on class interval selection was provided, and the choice of class interval looks at data for only one given time interval (or aggregate thereof). However, when the statistics are allowed to change temporally, the choice of class intervals becomes increasingly challenging. Momonier [104] noted that the placement of class breaks in animated choropleth maps magnifies the challenges of class choice. Now, the class choice must work not only across one map but across multiple maps,

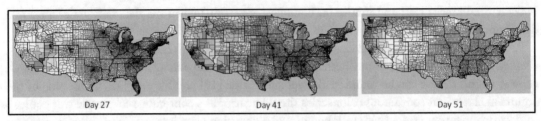

Figure 4.21: Animating time over choropleth maps. This figure shows the progression of a simulated pandemic over three time steps.

and the placement of these breaks can potentially emphasize relatively small fluctuations due to temporally global choices in class selection. Advanced classification schemes (e.g., [104]) have been developed to identify meaningful class breaks for time-series data; however, Harrower [64] argues that unclassified maps may be even more suitable for animation. Further details on designing effective animated maps can be found in work by Harrower [62, 63].

While the choice on class interval selection and the choice between classified and unclassified maps is left to the designer, it is clear that the addition of interaction techniques and animation (as shown in Griffin et al. [61]) can aid analysts in exploring their data. Overall, the addition of geospatial views for event exploration and spatiotemporal data can provide analysts with a powerful means of data exploration. However, the design choices in the visual parameters and the underlying data distributions must be adequately explored in order to avoid creating visualizations that can mislead or misinform the analyst.

4.8 SPATIAL ANOMALY DETECTION

While the use of interactive maps can aid an analyst in exploring their data, advanced data analysis methods can be employed to guide the analyst to areas of interest. As in temporal data, it is often the case that an analyst will want to find areas where the data appears to be statistically anomalous. In temporal data, it was noted that events that occur next to each other in time are likely to be related. For spatial data, a similar axiom holds, and it is denoted as Tobler's first law of geography [130]. Tobler's first law of geography states that "everything is related to everything else, but near things are more related than distant things." Choropleth maps let analysts visually identify regions that appear to be related; however, we can do analysis prior to the visualization to find areas that are statistically correlated and focus the visual representation on these areas.

4.8.1 SPATIAL AUTOCORRELATION

One method of analyzing the spatial relationships of variables is to explore the spatial autocorrelation between locations. If correlation exists (either positively or negatively), there are typically three explanations that would need to be explored by the analyst. The first explanation is that whatever is causing observations in one area is causing these to occur in another. For example, if a disease outbreak is occurring in several neighboring counties, and these counties all show a high spatial autocorrelation, the analyst might explore other data sets to see if there is an underlying commonality between these locations causing these observations. The second explanation is similar to the first, except that instead of an underlying commonality, there is a point source location causing events to occur, and these events are spreading out to neighboring regions. Finally, one could explore whether it is the intermingling of the populations that is causing events to spread. Often times, one or more of these explanations could play a role in the underlying analysis.

In order to determine if regions are spatially autocorrelated, there are two common statistical measures. The first common statistical measure is Moran's I [106]. In statistics, Moran's I is a measure of spatial autocorrelation, and it is characterized by a correlation in a signal among nearby locations in space. Moran's I is defined as:

$$I = \frac{N \sum_{i=1}^{N} \sum_{j=1}^{N} w_{ij}(X_i - \bar{X})(X_j - \bar{X})}{(\sum_{i=1}^{N} \sum_{j=1}^{N} w_{ij})(\sum_{i=1}^{N}(X_i - \bar{X})^2)} \quad (4.37)$$

In this equation, there is a set of N spatial units indexed by i and j, where X is the statistical measure of interest, $\bar{X}$ is the mean of X, and w_{ij} is the matrix of spatial weights which indicate the proximity of the neighboring statistics. Thus, given a set of global measures, Moran's I evaluates whether the pattern being analyzed is clustered, dispersed or random. Moran's I is bounded in the range of $[-1, 1]$ where values near 1 indicate clustering (high spatial autocorrelation), values near -1 indicate dispersion, and values near 0 indicate no relationship between the observations.

The second common statistical measure for calculating spatial autocorrelation is Geary's C.

$$C = \frac{(N-1) \sum_{i=1}^{N} \sum_{j=1}^{N} w_{ij}(X_i - X_j)^2}{2(\sum_{i=1}^{N} \sum_{j=1}^{N} w_{ij})(\sum_{i=1}^{N}(X_i - \bar{X})^2)} \quad (4.38)$$

As in Moran's I, there is a set of N spatial units indexed by i and j, where X is the statistical measure of interest, $\bar{X}$ is the mean of X, and w_{ij} is the matrix of spatial weights which indicate the proximity of the neighboring statistics. Geary's C values range from $[0, 3]$, which often makes them confusing to interpret; however, in comparison to Moran's I, Geray's C is more sensitive to extreme values and clustering. Values of C near 0 indicate clustering (high spatial autocorrelation), values near 1 indicate that there is no relationship between the observations, and values greater than 1 indicate dispersion.

4.8.2   LOCAL INDICATORS OF SPATIAL ASSOCIATION

However, both Moran's I and Geary's C look at a global spatial analysis and yield only one statistical measure of the entire data sample. Often analysts are interested in not only the global analysis of the data but also a local feature analysis. While no global autocorrelation may be present within the data set, it is still possible to find clusters that have local spatial autocorrelation. One such method of exploring local spatial autocorrelation is the local indicators of spatial association (LISA) work by

Anselin [9]. LISA utilizes the fact that Moran's I is a summation of individual cross products, and it evaluates the clustering across individual units by calculating Local Moran's I for each spatial unit and evaluating the statistical significance for each local I. The local indicators of spatial association are defined as:

$$I_i = \frac{N(X_i - \bar{X})}{\sum_{k=1}^{N}(X_k - \bar{X})^2} \sum_{j=1}^{N} w_{ij}(X_j - \bar{X})$$ (4.39)

As in the global Moran's I, there is a set of N spatial units indexed by i and j, where X is the statistical measure of interest, $\bar{X}$ is the mean of X, and w_{ij} is the matrix of spatial weights which indicate the proximity of the neighboring statistics. Now, each spatial location has its own local value of spatial autocorrelation, which can be used to indicate how related neighboring regions are.

4.8.3   AMOEBA CLUSTERING

Another method that utilizes local spatial statistics is the AMOEBA algorithm developed by Aldstadt and Getis [2]. The AMOEBA (A Multidirectional Optimum Ecotope-Based Algorithm) procedure is designed to identify local correlations in mapped data by assessing the spatial association of a particular mapped unit to its surrounding units. AMOEBA maps clusters of high and low values by creating a spatial weights matrix based on the Getis-Ord G_i^* statistic [60]. For a given location i, G_i^* is defined as

$$G_i^* = \frac{\sum_{j=1}^{N} w_{ij} x_j - \bar{x} \sum_{j=1}^{N} w_{ij}}{S\sqrt{\frac{[\sum N_{j=1} w_{if}^2 - (\sum_{j=1}^{N} w_{ij})^2]}{N-1}}}$$ (4.40)

Here, N is the number of spatial units, x_j is the value of interest within the areal unit at location j, $\bar{x}$ is the mean of all values, and

$$S = \sqrt{\frac{\sum_{j=1}^{N} x_j^2}{N} - (\bar{x}^2)}$$ (4.41)

w_{ij} is used as an indicator function that is one if j is a neighbor of i and zero, otherwise.

The AMOEBA algorithm develops a cluster from a selected seed location by evaluating G_i^* at all locations surrounding this seed location, and if the addition of a neighbor to the cluster increases the G_i^* value, then the neighbor is added. Details of this algorithm and the use of it in other visualization applications can be found in [2] and [76].

Figure 4.22 illustrates the application of AMOEBA clustering as applied to population statistics across the state of Indiana. Figure 4.22 (Left) shows a choropleth map of the percentage of the population under 18 in Indiana counties. Figure 4.22 (Right) shows the results of an AMOEBA clustering. Groups are colored based on their G_i^* values, and counties that connect to other counties of the same color are considered to be a cluster.

By utilizing measures of spatial autocorrelation and grouping, one can enhance the resulting data visualizations through different highlighting techniques or view point settings that guide the

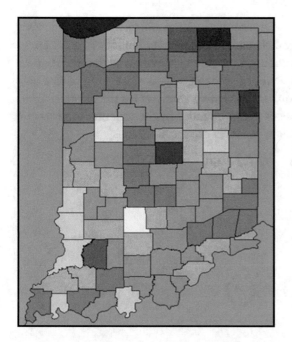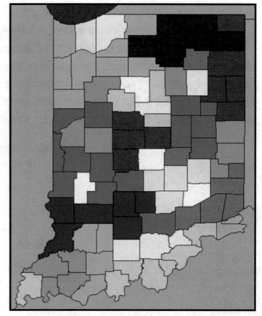

Figure 4.22: Aggregating data on population groups under 18. (Left) Choropleth map of the percent of population under 18. (Right) AMOEBA clustering of population based on percent under 18.

user to look at anomalous areas. Such statistics can guide the user in confirming their hypotheses about variable correlations across space while providing them with new insights into previously unknown correlations.

4.8.4   SPATIAL SCAN STATISTICS

While spatial correlations can guide users in forming hypotheses, the location of statistically anomalous clusters in space and time also provides insight into data sets. Detecting such clusters, or "hotspots," in data has become widespread with applications in fields such as epidemiology, data mining, astronomy, bio-surveillance, forestry, and uranium mining among others. Amongst the many methods for discovering hotspots are the geographical analysis machine by Openshaw[110] and the spatial scan statistic [90] developed by Kulldorff. Of the two methods, the spatial scan statistic has been the most widely adopted.

Kulldorff [90] proposed a *spatial scan statistic* to detect the location and size of the most likely cluster of events in spatial or spatiotemporal data. Using multidimensional data and a varying sized scanning window, the spatial scan statistic will give the location and size of statistically significant clusters of events when compared to a given statistical distribution of events of inhomogeneous density. Here, examples focus only on the spatial scan statistic with a Bernoulli distribution. A detailed description of scan statistics for other underlying distributions can be found in [58, 74, 82, 91].

Given a data set in the interval $[a, b]$, bounding a number of randomly placed points, such as illustrated in Figure 4.23, the algorithm first defines a scanning window $[t, t + w]$ of size $w <$ $(b - a)$. Figure 4.23 (Top) shows a number of different scanning windows for the illustrated data set. The scan statistic S_w is the maximum number of points bounded by the scanning window as it slides along the interval. Let $N(A)$ be the number of points in the set A. Then,

$$S_w = \sup_{a < t < (b - w)} N[t, t + w]$$

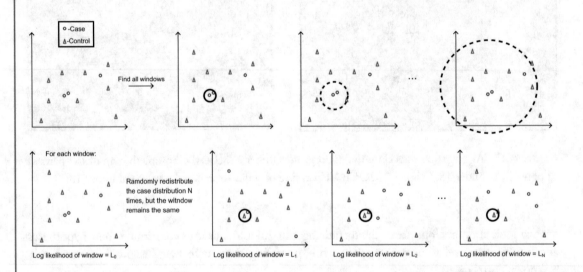

Figure 4.23: An illustration of determining window size and the log likelihood calculation steps in the spatial scan statistic. (Top) For each case in the data set, the algorithm iteratively finds all circular windows centered at the case. (Bottom) For each window, a log likelihood value is calculated. The data is redistributed N times and each time, the log likelihood of the window is recalculated.

As the scanning window W of variable size and shape is moved across the 2D area of interest $G \subset V$ for some vector space V, it defines a set of windows $\mathcal{W}$. The probability of an event within a window W is p and the probability of an event outside a window is q. The variable scanning window is illustrated in Figure 4.23 (Top). Under the null hypothesis, H_0, $p = q$. The alternative hypothesis, H_1 is that there is a window W such that $p > q$.

For the Bernoulli distribution model, each entity in G is in one of two states (either a case or a control as seen in Figure 4.23). Thus, for any subset $A \subset G$, A follows a binomial distribution. Under the null hypothesis, $N(A) \sim \text{Bin}(\mu(A), p)$. Under the alternative hypothesis, $N(A) \sim \text{Bin}(\mu(A), p)$ for all sets $A \subset W$ and $N(A) \sim \text{Bin}(\mu(A), q)$ for all sets $A \subset W^C$.

For each possible window W in the set of all windows $\mathcal{W}$, a likelihood value $L(W)$ is calculated based on the contents of the window, Figure 4.23 (Bottom). The likelihood value is maximized over all possible windows and this maximum likelihood is called the scan statistic λ. n_W and n_G are the number of events in a window W and area G, respectively. $\mu(G)$ and $\mu(W)$ are total number of points in area G and W, respectively.

$$p = \frac{n_W}{\mu(W)}$$

$$q = \frac{n_G - n_W}{\mu(G) - \mu(W)}$$

$$L_0 = \left(\frac{n_G}{\mu(G)}\right)^{n_G} \left(\frac{\mu(G) - n_G}{\mu(G)}\right)^{\mu(G) - n_G}$$

$$L(w) = \begin{cases} p^{n_W}(1-p)^{\mu(W)-n_W} q^{n_G - n_W} \\ \quad (1-q)^{(\mu(G)-\mu(W))-(n_G - n_W)} & \text{if } p > q, \\ L_0 & \text{otherwise} \end{cases}$$

$$\lambda = \max_{w \in \mathcal{W}} \frac{L(w)}{L_0} \tag{4.42}$$

An analytical description of the distribution of the test statistic does not exist. Thus, a Monte Carlo simulation [44] is used to obtain the distribution of the test statistic under the null hypothesis. As such, for each window, the data is randomly redistributed (Figure 4.23), and the likelihood value for a given window is calculated under each new distribution. The test statistic for the actual data is then compared to the distribution of the test statistic under the null hypothesis to reject or not reject the null hypothesis with α significance. In other words, once the N likelihood values are calculated, they are sorted in descending order, and the p-value is calculated as the position of λ in the list divided by $N + 1$. Finally, when a likelihood value and significance value have been computed for every possible window, a list of scan windows based on location, radius, likelihood and significance are returned.

Such work is an underpinning in visual analytics as interactive data exploration methods that can incorporate end user knowledge and enable users to test hypotheses about whether a given subset of their data is truly a statistical anomaly. By combining the user's expert knowledge, the data space on which the time intensive scan statistic is run can be reduced. The output of spatial scan statistics have been used in a variety of applications (e.g., [23]), and the resulting visualizations of such clusters aids the end user in finding statistically significant regions within their dataset.

CHAPTER 5

Summary

As shown throughout this chapter, there is no single catch-all visual representation or analysis. Different analyses and representations allow us to search for different components, discover different features, and analyze data in a variety of unique ways. Given these different representations and analyses, one can begin combining these techniques to create even more comprehensive tools and interactive visuals that can further aid the analysis process.

By combining analysis and data preconditioning as a part of the visualization pipeline, one can potentially create more effective visuals. The combination of such visuals with interactive feedback for adjusting algorithm parameters and further exploring details within a dataset is an equally important step within the visual analytics process. This interactivity can lead to new insights and pose new problems in which analytical algorithms may need to be applied again. This process of teasing out information needs to loop back to the underlying analytics as a means of evaluation and hypothesis testing.

As previously stated, we need to be cognizant of design parameters not only in the modeling and analysis of the data but also of design parameters for visual representations. All visualization is subject to design decisions and many conflicting requirements. The tools presented in this chapter have their own strengths and weaknesses, and as the amount of data being stored and process increases, the need for tools to facilitate the analytical process is ever increasing. Currently, a variety of software packages (i. e. MatLab, R and Mondrian [129]) exist in which implementations of the described techniques can be further explored.

Bibliography

[1] W. Aigner, S. Miksch, W. Muller, H. Schumann and C. Tominski, "Visual methods for analyzing time-oriented data," *IEEE Transactions on Visualization and Computer Graphics*, vol. 14, no. 1, pp. 47–60, 2008. DOI: 10.1109/TVCG.2007.70415

[2] J. Aldstadt and A. Getis, "Using AMOEBA to create a spatial weights matrix and identify spatial clusters," *Geographical Analysis*, vol. 38, pp. 327–343, 2006. DOI: 10.1111/j.1538-4632.2006.00689.x 56

[3] D. F. Andrews, "Plots of High-Dimensional Data," *Biometrics*, March, pp. 125–136, 1972. DOI: 10.2307/2528964

[4] N. Andrienko, G. Andrienko and P. Gatalsky, "Exploratory Spatio-Temporal Visualization: An Analytical Review," *Journal of Visual Languages & Computing*, vol. 14, no. 6, pp. 503–541, 2003. DOI: 10.1016/S1045-926X(03)00046-6 49

[5] G. Andrienko, D. Malerba, M. May and M. Teisseire, "Mining spatio-temporal data," *Journal of Intelligent Information Systems*, vol. 27, pp. 187–190, 2006. DOI: 10.1007/s10844-006-9949-3 49

[6] N. Andrienko and G. Andrienko, "Interactive Maps for Visual Data Exploration," *International Journal of Geographic Information Science*, vol. 13, no. 4, pp. 355–374, 1999. DOI: 10.1080/136588199241247 49

[7] N. Andrienko and G. Andrienko, *Exploratory Analysis of Spatial and Temporal Data: A Systematic Approach*, Springer-Verlag, 2006. 49

[8] M. Ankerst, S. Berchtold and D. A. Keim, "Similarity clustering of dimensions for an enhanced visualization of multidimensional data," *IEEE Symposium on Information Visualization*, pp. 52–62, 1998. DOI: 10.1109/INFVIS.1998.729559 28

[9] L. Anselin, "Local Indicators of Spatial Association - LISA," *Geographical Analysis*, vol. 27, pp. 93–115, 1995. DOI: 10.1111/j.1538-4632.1995.tb00338.x 56

[10] R. W. Armstrong, "Standardized Class Intervals and Rate Computation in Statistical Maps of Mortality," *Annals of the Association of American Geographers*, vol. 59, no. 2, pp. 382–390, 1969. DOI: 10.1111/j.1467-8306.1969.tb00677.x 52

[11] M. P. Armstrong, N. C. Xiao and D. A. Bennett, "Using Genetic Algorithms to Create Multicriteria Class Intervals for Choropleth Maps," *Annals of the Association of American Geographers*, vol. 93, no. 3, pp. 595–623, 2003. DOI: 10.1111/1467-8306.9303005 52

[12] S. Bachthaler and D. Weiskopf, "Continuous Scatterplots," *IEEE Transactions on Visualization and Computer Graphics*, vol. 14, no. 6, pp. 1428–1435, 2008. DOI: 10.1109/TVCG.2008.119 27

[13] R. A. Becker and W. S. Cleveland, "Brushing Scatterplots," *Technometrics*, vol. 29, no. 2, pp. 127–142, 1987. DOI: 10.2307/1269768 28

[14] R. A. Becker, W. S. Cleveland and M.-J. Shyu, "The Visual Design and Control of Trellis Display," *Journal of Computational Graphical Statistics*, vol. 5, no. 2, pp. 123–155, 1996. DOI: 10.2307/1390777 41

[15] D. Borland and R. M. Taylor, "Rainbow Color Map (Still) Considered Harmful," *Computer Graphics & Applications*, vol. 27, no. 2, pp. 14–17, 2007. DOI: 10.1109/MCG.2007.323435 6

[16] G. Box and D. Cox, "An analysis of transformations," *Journal of the Royal Statistical Society, Series B (Methodological)*, vol. 26, no. 2, pp. 211–252, 1964. 11, 12, 13

[17] G. Box and G. Jenkins, *Time series analysis: Forecasting and control*, San Francisco: Holden-Day, 1970. 45

[18] C. A. Brewer, "Designing Better Maps: A Guide for GIS Users," *ESRI Press*, 2005. 53

[19] C. A. Brewer and L. Pickle, "Evaluation of Methods for Classifying Epidemiological Data on Choropleth Maps in Series," *Annals of the Association of American Geographers*, vol. 92, no. 4, pp. 662–681, 2002. DOI: 10.1111/1467-8306.00310 11, 52, 53

[20] P. Buono, C. Plaisant, A. Simeone, A. Aris, G. Shmueli and W. Jank, "Similarity-Based Forecasting with Simultaneous Previews: A River Plot Interface for Time Series Forecasting," *Proceedings of the 11th International Conference on Information Visualization*, 2007. DOI: 10.1109/IV.2007.101

[21] J. V. Carlis and J. A. Konstan, "Interactive Visualization of Serial Periodic Data," *Proceedings of the Symposium on User Interface Software and Technology (UIST)*, 1998. DOI: 10.1145/288392.288399

[22] D. B. Carr, D. White and A. M. MacEachren, "Conditioned Choropleth Maps and Hypothesis Generation," *Annals of the Association of American Geographers*, vol. 95, no. 1, pp. 32–53. DOI: 10.1111/j.1467-8306.2005.00449.x 49

[23] J. Chen, R. E. Roth, A. T. Naito, E. J. Lengerich and A. M. MacEachren, "Geovisual Analytics to Enhance Spatial Scan Statistic Interpretation: An Analysis of U. S. Cervical Cancer Mortality," *International Journal of Health Geographics*, vol. 7, no. 57, 2008. DOI: 10.1186/1476-072X-7-57 59

[24] H. Chernoff, "The use of faces to represent points in k-dimensional space graphically," *Journal of the American Statistical Association*, vol. 68, pp. 361–368, 1973. DOI: 10.2307/2284077 31

[25] E. H. Chi, J. Riedl, B. Phillip and J. Konstan, "Principles for Information Visualization Spreadsheets," *IEEE Computer Graphics and Applications*, vol. 18, no. 4, pp. 30–38, 1998. DOI: 10.1109/38.689659

[26] E. H. Chi and S. K. Card, "Sensemaking of Evolving Web Sites Using Visualization Spreadsheets," *IEEE Symposium on Information Visualization*, pp. 18–27, 1999.

[27] L. Chittaro and C. Combi, "Visualizing Queries on Databases of Temporal Histories: New Metaphors and Their Evaluation," *Proceedings of the IEEE Symposium on Information Visualization*, 2001. DOI: 10.1016/S0169-023X(02)00137-4

[28] J. Choo, Hanseung Lee, J. Kihm and H. Park, "iVisClassifier: An Interactive Visual Analytics System for Classification Based on Supervised Dimension Reduction," *Proceedings of the IEEE Conference on Visual Analytics*, pp. 27–34, 2010. DOI: 10.1109/VAST.2010.5652443 35

[29] W. S. Cleveland and R. McGill, "Graphical Perception: Theory, Experimentation and Application to the Development of Graphical Methods," *Journal of American Statistical Association*, vol. 79, no. 387, pp. 531–554, 1984. DOI: 10.2307/2288400 11

[30] W. S. Cleveland and R. McGill, "Graphical Perception and Graphical Methods for Analyzing Scientific Data," *Science*, vol. 30, pp. 828–833, 1985. DOI: 10.1126/science.229.4716.828 11

[31] W. S. Cleveland, *The Elements of Graphing Data*, Wadsworth Publishing Company, 1985. 11

[32] W. S. Cleveland, "Dynamic Graphics for Statistics," *Statistics/Probability Series*, Wadsworth& Brooks/Cole, 1988.

[33] W. S. Cleveland and S. J. Devlin, "Locally-Weighted Regression: An Approach to Regression Analysis by Local Fitting," *Journal of American Statistical Association*, vol. 83, pp. 596–610, 1988. DOI: 10.2307/2289282 46

[34] R. B. Cleveland, W. S. Cleveland, J. E. McRae and I. Terpenning, "STL: A Seasonal-Trend Decomposition Procedure Based on Loess," *Journal of Official Statistics*, vol. 6, pp. 3–73, 1990. 46, 47

[35] W. S. Cleveland, *Visualizing Data*, Hobart Press, 1993. 11, 12, 13, 26, 28

[36] W. S. Cleveland, "A Model for Studying Display Methods of Statistical Graphics," *Journal of Computational and Graphical Statistics*, vol. 2, no. 4, pp. 323–364, 1993. DOI: 10.2307/1390686 11

[37] R. D. Cook and S. Weisberg, *Applied Regression Including Computing and Graphics*, John Wiley, 1999. 14

[38] N. Cressie, *Statistics for Spatial Data*, Wiley, New York, 1993. 51

[39] E. K. Cromley and R. G. Cromley, "An analysis of alternative classifications schemes for medical atlas mapping," *European Journal of Cancer, Series B (Methodological)*, vol. 26, no. 2, pp. 211–252, 1964. DOI: 10.1016/0959-8049(96)00130-X 52, 53

[40] T. N. Dang, L. Wilkinson and A. Anand, "Stacking Graphic Elements to Avoid Over-Plotting," *IEEE Transactions on Visualization and Computer Graphics*, vol. 14, no. 6, pp. 1044–1052, 2010. DOI: 10.1109/TVCG.2010.197 21

[41] C. Daniel, F. S. Wood and J. W. Gorman, *Fitting Equations to Data*, Wiley New York, 1980. 11

[42] L. Denby and C. Mallows, "Variations of the Histogram," *Journal of Computational and Graphical Statistics*, vol. 18, no. 1, pp. 21–31, 2009. DOI: 10.1198/jcgs.2009.0002 19

[43] C. Ding and X. He, "K-means Clustering via Principal Component Analysis," *Proceedings of the International Conference on Machine Learning*, pp. 225–232, 2004. DOI: 10.1145/1015330.1015408 35

[44] M. Dwass, "Modified Randomization Tests for Nonparametric Hypotheses," *The Annals of Mathematical Statistics*, vol. 28, no. 1, pp. 181–187, 1957. DOI: 10.1214/aoms/1177707045 59

[45] J. A. Dykes, "Exploring Spatial Data Representation with Dynamic Graphics," *Computational Geoscience*, vol. 23, no. 4, pp. 345–370, 1997. DOI: 10.1016/S0098-3004(97)00009-5 49

[46] J. Dykes and A. M. MacEachren, *Exploring Geovisualization*, Elsevier, 2005. 49

[47] J. A. Dykes and D. M. Mountain, "Seeking Structure in Records of Spatio-Temporal Behaviour: Visualization Issues, Efforts and Applications," *Computational Statistics and Data Analysis*, vol. 43, no. 4, pp. 581–603, 2003. DOI: 10.1016/S0167-9473(02)00294-3 49

[48] J. A. Dykes and D. J. Unwin, "Maps of the Census: A Rough Guide," *Case Studies of Visualization in the Social Sciences: Technical Report*, vol. 43, (Eds, D.J. Unwin and P.F. Fisher), Loughborough, UK: Advisory Group on Computer Graphics, 29-54, 2003. 53

[49] R. M. Edsall, M. Harrower and J. L. Mennis, "Tools for visualizing properties of spatial and temporal periodicity in geographic data," *Computational Geoscience*, vol. 26, no. 1, pp. 109–118, 2000. DOI: 10.1016/S0098-3004(99)00037-0 49

[50] C. L. Eicher and C. A. Brewer, "Dasymetric Mapping and Areal Interpolation: Implementation and Evaluation," *Cartography and Geographic Information Science*, pp. 125–138, 2001. DOI: 10.1559/152304001782173727 51

[51] N. Elmqvist, P. Dragicevic, and J.-D. Fekete, "Rolling the Dice: Multidimensional Visual Exploration Using Scatterplot Matrix Navigation," *IEEE Transactions on Visualization and Computer Graphics*, vol. 14, no. 6, pp. 1141–1148, 2008. DOI: 10.1109/TVCG.2008.153 28

[52] N. Elmqvist, P. Dragicevic, and J.-D. Fekete, "Color Lens: Adaptive Color Scale Optimization for Visual Exploration," *IEEE Transactions on Visualization and Computer Graphics*, To Appear, 2010. DOI: 10.1109/TVCG.2010.94 9

[53] S. E. Fienberg, "Graphical Methods in Statistics," *The American Statistician*, vol. 33, pp. 156–178, 1979. DOI: 10.2307/2683729 31

[54] S. Few, *Now You See It: Simple Visualization Techniques for Quantitative Analysis*, Analytics Press, 2009. 43

[55] B. Flury and H. Riedwyl, "Graphical Representation of Multivariate Data by Menas of Asymmetrical Faces," *Journal of the American Statistical Association*, vol. 76, pp. 757–765, 1981. DOI: 10.2307/2287565 31

[56] A. U. Frank, "Different Types of 'Times' in GIS," *Spatial Temporal Reasoning in Geographic Information Systems*, Oxford University Press, 1998.

[57] D. A. Freedman, *Statistical Models: Theory and Practice*, Cambridge University Press, 2005.

[58] J. Glaz, J. Naus and S. Wallenstein, *Scan Statistics*, Springer-Verlag, 2001. 58

[59] R. C. Geary, "The Contiguity Ratio and Statistical Mapping," *The Incorporated Statistician*, vol. 5, no. 3, pp. 115–145, 1954. DOI: 10.2307/2986645

[60] A. Getis and J. K. Ord, "The Analysis of Spatial Association by Use of Distance Statistics," *Geographical Analysis*, vol. 24, pp. 189–206, 1992. DOI: 10.1111/j.1538-4632.1992.tb00261.x 56

[61] A. Griffin, A. MacEachren, F. Hardisty, E. Steiner and B. Li, "A Comparison of Animated Maps with Static Small-Multiple Maps for Visually Identifying Space-Time Clusters," *Annals of the Association of American Geographers*, vol. 96, no. 4, pp. 740–753, 2006. DOI: 10.1111/j.1467-8306.2006.00514.x 42, 54

[Hardin and Maffi] C. Hardin and L. Maffi, *Color Categories in Thought and Language*, Cambridge University Press, 1997. 7

[62] M. A. Harrower, "Tips for Designing Effective Animated Maps," *Cartographic Perspectives*, vol. 44, pp. 63–65, 2003. 54

[63] M. A. Harrower, "A Look at the History and Future of Animated Maps," *Cartographica: The International Journal for Geographic Information and Geovisualization*, vol. 39, no. 3, pp. 33–42, 2004. DOI: 10.3138/7MN7-5132-1MW6-4V62 54

[64] M. A. Harrower, "Unclassed Animated Choropleth Maps," *The Cartographic Journal*, vol. 44, no. 4, pp. 313–320, 2007. DOI: 10.1179/000870407X241863 53, 54

[65] M. A. Harrower and C. A. Brewer, "ColorBrewer.org: An online tool for selecting color schemes for maps," *The Cartographic Journal*, vol. 40, no. 1, pp. 27–37, 2003. DOI: 10.1179/000870403235002042 6

[66] J. A. Hartigan, *Clustering Algorithms*, Wiley, 1975. 34

[67] W. W. Hargrove and F. M. Hoffman, "Using Multivariate Clustering to Characterize Ecoregion Borders," *Computing in Science & Engineering*, vol. 1, no. 4, pp. 18–25, 1999. DOI: 10.1109/5992.774837 35, 49

[68] H. Hauser, F. Ledermann and H. Doleisch, "Angular Brushing of Extended Parallel Coordinates," *Proceedings of the IEEE Symposium on Information Visualization*, pp. 127–135, 2002. DOI: 10.1109/INFVIS.2002.1173157 11

[69] S. Havre, E. Hetzler, P. Whitney and L. Nowell, "Themeriver: Visualizing thematic changes in large document collections," *IEEE Transactions on Visualization and Computer Graphics*, vol. 8, no. 1, pp. 9–20, 2002. DOI: 10.1109/2945.981848 39

[70] J. Heer and N. Kong and M. Agrawala, "Sizing the Horizon: The Effects of Chart Size and Layering on the Graphical Perception of Time Series Visualization," *Proceedings of the 27th international conference on Human factors in computing systems*, pp. 1303–1312, 2009. DOI: 10.1145/1518701.1518897 11

[71] J. Heinrich and D. Weiskopf, "Continuous Parallel Coordinates," *IEEE Transactions on Visualization and Computer Graphics*, vol. 15, no. 6, pp. 1531–1538, 2009. DOI: 10.1109/TVCG.2009.131 30

[72] K. P. Hewagamage, M. Hirakawa and T. Ichikawa, "Interactive Visualization of Spatiotemporal Patterns Using Spirals on a Geographical Map," *Proceedings of the IEEE Symposium on Visual Languages*, 1999. DOI: 10.1109/VL.1999.795916 40

[73] H. Hochheiser and B. Shneiderman, "Dynamic Query Tools for Time Series Data Sets: Timebox Widgets for Interactive Exploration," *Information Visualization*, vol. 3, no. 1, pp. 1–18, 2009. DOI: 10.1057/palgrave.ivs.9500061

[74] L. Huang, M. Kulldorf and A. Klassen, "A Spatial Scan Statistic for Survival Data," *Biometrics*, vol. 63, pp. 109–118, 2007. DOI: 10.1111/j.1541-0420.2006.00661.x 58

[75] A. Inselberg, "The Plane with Parallel Coordinates," *The Visual Computer*, vol. 1, no. 2, pp. 69–91, 1985. DOI: 10.1007/BF01898350 29

[76] M. Jankowska, J. Aldstadt, A. Getis, J. Weeks and G. Fraley, "An AMOEBA Procedure for Visualizing Clusters," *Proceedings of GIScience 2008*, 2008. 56

[77] T. J. Jankun-Kelly and K.-L. Ma, "Visualization, Exploration and Encapsulation via a Spreadsheet-Like Interface," *IEEE Transactions on Visualization and Computer Graphics*, vol. 7, no. 3, pp. 275–287, 2001. DOI: 10.1109/2945.942695

[78] G. F. Jenks, "The data model concept in statistical mapping," *International Yearbook of Cartography*, vol. 26, pp. 186–190, 1964. 52, 53

[79] D. H. Jeong, C. Ziemkiewicz, B. Fisher, W. Ribarsky and R. Chang, "iPCS: An Interactive System for PCA-based Visual Analytics," *Computer Graphics Forum*, vol. 28, no. 3, pp. 767–774, 2009. DOI: 10.1111/j.1467-8659.2009.01475.x 35

[80] N. K. Jog and B. Shneiderman, "Starfield Visualization with Interactive Smooth Zooming," *Proceedings of the Third IFIP WG2.6 Working Conference on Visual Database Systems 3 (VDB-3)*, pp. 3–14, 1995.

[81] I. T. Jolliffe, "Principal Component Analysis," *Springer Series in Statistics*, Springer, NY, 2002. DOI: 10.1002/0470013192.bsa501 32

[82] I. Jung, M. Kulldorff, and A. Klassen, "A Spatial Scan Statistic for Ordinal Data," *Statistics in Medicine*, vol. 26, pp. 1594–1607, 2007. DOI: 10.1002/sim.2607 58

[83] D. A. Keim, "Designing Pixel-Oriented Visualization Techniques: Theory and Applications," *IEEE Transactions on Visualization and Computer Graphics*, vol. 6, no. 1, pp. 59–78, 2000. DOI: 10.1109/2945.841121 31

[84] D. A. Keim, "Information visualization and visual data mining," *IEEE Transactions on Visualization and Computer Graphics*, vol. 8, no. 1, pp. 1–8, 2002. DOI: 10.1109/2945.981847 28

[85] P. Kidwell, G. Lebanon and W. S. Cleveland, "Visualizing Incomplete and Partially Ranked Data," *IEEE Transactions on Visualization and Computer Graphics*, vol. 14, pp. 1356–1363, 2008. DOI: 10.1109/TVCG.2008.181 11

[86] T. Kohonen, "The self-organizing map," *Proceedings of the IEEE*, pp. 1464–1480, 1986. 37

[87] Y. Koren and L. Carmel, "Robust linear dimensionality reduction," *IEEE Transactions on Visualization and Computer Graphics*, vol. 10, no. 4, pp. 459–470, 2004. DOI: 10.1109/TVCG.2004.17 34

[88] R. Kosara, F. Bendix, and H. Hauser, "Parallel Sets: Interactive Exploration and Visual Analysis of Categorical Data," *IEEE Transactions on Visualization and Computer Graphics*, vol. 12, no. 4, pp. 558–568, 2006. DOI: 10.1109/TVCG.2006.76 30

[89] J. B. Kruskal and M. Wish, *Multidimensional Scaling*, Sage Publications, 1978. 36

[90] M. Kulldorf, "A spatial scan statistic," *Communications in Statistics: Theory and Methods*, vol. 26, pp. 1481–1496, 1997. DOI: 10.1145/1150402.1150410 57

[91] M. Kulldorf, L. Huang, and K. Konty, "A Scan Statistic for Continuous Data Based on the Normal Probability Model," *International Journal of Health Geographics*, vol. 8, no. 58, 2009. DOI: 10.1186/1476-072X-8-58 58

[92] R. J. Larsen and M. L. Marx, *An Introduction to Mathematical Statistica and Its Applications*, Prentice Hall, 2000.

[93] H. Levkowitz and G. T. Herman, "Color scales for image data," *IEEE Computer Graphics and Applications*, vol. 12, pp. 72–80, 1992. DOI: 10.1109/38.135886 5

[94] M. Levoy, "Spreadsheets for Images," *Proceedings of the 21st Annual Conference on Computer Graphics and Interactive Techniques (SIGGRAPH)*, pp. 139–146, 1994. DOI: 10.1145/192161.192190

[95] R. J. Littlefield, "Using the glyph concept to create user-definable display formats," *Proceedings of the NCGA*, pp. 697–706, 1983. 30

[96] A. MacEachren, *How Maps Work*, Guilford Press, 1995. 49

[97] A. MacEachren, "Visualizing Uncertain Information," *Cartographic Perspective*, vol. 9, no. 3, pp. 378–394, 1992. 8

[98] J. B. MacQueen, "Some methods for classification and analysis of multivariate observations," *Proceedings of the 5th Berkeley Symposium on Mathematical Statistics and Probability*, pp. 281–297, 1967. 34

[99] R. Maciejewski, R. Hafen, S. Rudolph, G. Tebbetts, W. S. Cleveland, S. J. Grannis, and D. S. Ebert, "Generating Synthetic Syndromic-Surveillance Data for Evaluating Visual-Analytics Techniques," *IEEE Computer Graphics and Applications*, vol. 29, no. 3, pp. 18–28, 2009. DOI: 10.1109/MCG.2009.43 14

[100] J. R. Mackay, "An Analysis of Isopleth and Choropleth Class Intervals," *Economic Geography*, vol. 31, pp. 71–81, 1955. 52

[101] A. Mead, "Review of the development of multidimensional scaling methods," *The Statistician*, vol. 3, pp. 27–35, 1992. 36

[102] J. R. Miller, "Attribute Blocks: Visualizing Multiple Continuously Defined Attributes," *IEEE Computer Graphics and Applications*, vol. 27, no. 3, pp. 57–69, 2007. DOI: 10.1109/MCG.2007.54 8

[103] M. S. Monmonier, "Contiguity-Biased Class-Interval Selection: A Method for Simplifying Patterns on Statistical Maps," *Geographical Review*, vol. 62, no. 2, pp. 203–228, 1972. 11, 52

[104] M. S. Monmonier, "Minimum-change categories for dynamic temporal choropleth maps," *Journal of the Pennsylvania Academy of Science*, vol. 68, no. 1, pp. 42–47, 1994. 53, 54

[105] D. Montgomery, *Introduction to Statistical Quality Control*, John Wiley & Sons, 2005. 44

[106] P. A. P. Moran, "Notes on continuous stochastic phenomena," *Biometrika*, vol. 37, pp. 17–33, 1950. 55

[107] K. Moreland, "Diverging Color Maps for Scientific Visualization," *Proceedings of the 5th International Symposium on Visual Computing*, December 2009. DOI: 10.1007/978-3-642-10520-3_9 5

[108] C. J. Morris, D. S. Ebert, and P. Rheingans, "An Experimental Analysis of the Pre-Attentiveness of Features in Chernoff Faces," *SPIE Proceedings of Applied Imagery Pattern Recognition: 3D Visualization for Data Exploration and Decision Making*, 1999. 31

[109] W. Müller, T. Nocke and H. Schumann, "Enhancing the Visualization Process with Principal Component Analysis to Support the Exploration of Trends," *Proceedings of the 2006 Asia-Pacific Symposium on Information Visualization*, pp. 121–130, 2006.

[110] S. Openshaw, M. Charlton, C. Wymer and A. Craft, "A mark I geographical analysis machine for the automated analysis of point data sets," *International Journal of Geographical Information Systems*, vol. 1, pp. 335–358, 1987. DOI: 10.1080/02693798708927821 57

[111] K. Pearson, "Contributions to the Mathematical Theory of Evolution. II. Skew Variation in Homogenous Material," *Philosophical Transactions of the Royal Society A: Mathematical, Physical and Engineering Sciences*, vol. 186, pp. 326–343, 1895. DOI: 10.1098/rsta.1895.0010 17

[112] D. Peuquet, *Representations of Space and Time*, Guilford Press., 2002. 49

[113] B. Pham, "Spline-based Color Sequences for Univariate, Bivariate and Trivariate Mapping," *Proceedings of the IEEE Conference on Visualization*, pp. 208–202, 1990. DOI: 10.1109/VISUAL.1990.146383 8

[114] W. Peng, M. O. Ward, and E. A. Rundensteiner, "Clutter reduction in multi-dimensional data visualization using dimension reordering," *Proceedings of the IEEE Symposium on Information Visualization*, pp. 89–96, 2004. DOI: 10.1109/INFOVIS.2004.15

[115] P. Rheingans, "Task-based color scale design," *Proceedings of Applied Image and Pattern Recognition*, pp. 35–43, 1999. DOI: 10.1117/12.384882 5, 6, 7, 8

[116] G. Robertson, R. Fernandez, D. Fisher, B. Lee and J. Stasko, "Effectiveness of Animation in Trend Visualization," *IEEE Transactions on Visualization and Computer Graphics*, vol. 14, no. 6, pp. 1325–1332, 2008. DOI: 10.1109/TVCG.2008.125 43

[117] S. Schiffman, M. L. Reynolds and F. W. Young, *Introduction to Multidimensional Scaling: Theory, Methods, and Applications*, Academic Press, New York. DOI: 10.1016/j.datak.2010.12.003 36

[118] P. Schulze-Wollgast, C. Tominski, and H. Schumann, "Enhancing Visual Exploration by Appropriate Color Coding," *Proceedings of International Conference in Central Europe on Computer Graphics, Visualization and Computer Vision (WSCG)*, pp. 203–210, 2005. 52

[119] D. W. Scott, "On optimal and data-based histograms," *Biometrika*, vol. 66, no. 3, pp. 605–610, 1979. DOI: 10.1093/biomet/66.3.605 19

[120] W. A. Shewart, *Economic Control of Quality of Manufactured Product*, D. Van Nostrand Company, Inc., 1931. 43

[121] B. Shneiderman, "The eyes have it: A task by data type taxonomy for information visualizations," *Proceedings of the IEEE Symposium on Visual Languages*, pp. 336–343, 1996. DOI: 10.1109/VL.1996.545307 17, 28

[122] B. W. Silverman, *Density Estimation for Statistics and Data Analysis*, Chapman & Hall/CRC, 1986. 22, 23, 25

[123] T. A. Slocum, R. B. McMaster, F. C. Kessler and H. H. Howard, *Thematic Cartography and Geographic Visualization*, Prentice Hall, Englewood Cliffs, NJ, 2005. 53

[124] R. M. Smith, "Comparing Traditional Methods for Selecting Class Intervals on Choropleth Maps," *Professional Geographer*, vol. 38, no. 1, pp. 62–67, 1986. DOI: 10.1111/j.0033-0124.1986.00062.x 52

[125] S. S. Stevens, "On the Theory of Scales of Measurement," *Science*, vol. 103, no. 2684, pp. 677–680, 1946. DOI: 10.1126/science.103.2684.677 3

[126] M. A. Stoto and J. D. Emerson, "Power Transformations for Data Analysis," *Sociological Methodology*, vol. 14, pp. 126–168, 1983. DOI: 10.2307/270905 11

[127] H. A. Sturges, "The choice of a class interval," *Journal of the American Statistical Association*, vol. 21, no. 153, pp. 65–66, 1926. 18, 19

[128] J. J. Thomas and K. A. Cook, *Illuminating the Path: The R&D Agenda for Visual Analytics*, National Visualization and Analytics Center, 2005. 1

[129] M. Theus, "Interactive Data Visualization using Mondrian," *Journal of Statistical Software*, vol. 7, no. 11, pp. 1–9, 2002. DOI: 10.1002/wics.120 61

[130] W. Tobler, "A computer movie simulating urban growth in the Detroit region," *Economic Geography*, vol. 46, no. 2, pp. 234–240, 1970. DOI: 10.2307/143141 54

[131] C. Tominski, J. Abello and H. Schumann, "Axes-based visualizations with radial layouts," *Proceedings of the ACM Symposium on Applied Computing*, pp. 1242–1247, 2004. DOI: 10.1145/967900.968153 42

[132] B. E. Trumbo, "Theory for Coloring Bivariate Statistical Maps," *The American Statistician*, vol. 35, no. 4, pp. 220–226, 1981. DOI: 10.2307/2683294 5, 7

[133] E. R. Tufte, *The Visual Display of Quantitative Information*, Graphics Press, 1983. 26

[134] E. R. Tufte, *Envisioning Information*, Graphics Press, 1990. 41

[135] J. W. Tukey, "On the comparative anatomy of transformations," *Annals of Mathematical Statistics*, vol. 28, pp. 602–632, 1955. DOI: 10.1214/aoms/1177706875 11

[136] J. W. Tukey, *Exploratory Data Analysis*, University Microfilms International, 1988. 11

[137] J. J. van Wijk and E. R. van Selow, "On optimal and data-based histograms," *Biometrika*, vol. 66, no. 3, pp. 605–610, 1979. DOI: 10.1093/biomet/66.3.605 40

[138] C. Ware, *Information Visualization: Perception for Design*, Morgan Kaufmann, 2004. 4

[139] M. Weber, M. Alexa and W. Muller, "Visualizing time-series on spirals," *Proceedings of the IEEE Symposium on Information Visualization*, pp. 7–14, 2001. DOI: 10.1109/INFVIS.2001.963273 40

[140] L. Wilkinson, "Algorithms for Choosing the Domain and Range when Plotting a Function," *Computing and Graphics in Statistics*, pp. 231–237, 1991. 11

[141] L. Wilkinson, "Dot plots," *The American Statistician*, vol. 53, pp. 276–281, 1999. DOI: 10.2307/2686111 20

[142] L. Wilkinson, *The Grammar of Graphics*, Springer, 2005. 17

[143] J. K. Wright, "Problems in population mapping," *in Notes on Statistical Mapping, With Special Reference to the Mapping of Population Phenomena*, American Geographical Association of America, pp. 1–18, 1938. 50

[144] J. Wang, W. Peng, M. O. Ward, and E. A. Rundensteiner, "Interactive hierarchical dimension ordering, spacing and filtering for exploration of high dimensional datasets," *Proceedings of the IEEE Symposium on Information Visualization*, pp. 105–112, 2003. DOI: 10.1109/INFVIS.2003.1249015 30

[145] H. Zha, C. Ding, M. Gu, X. He and H. D. Simon, "Spectral Relaxation for K-Means Clustering," *Neural Information Processing Systems*, vol. 14, pp. 1057–1064, 2001. 35

Author's Biography

ROSS MACIEJEWSKI

Ross Maciejewski received his PhD in 2009 from Purdue University for his thesis "Exploring Multivariate Data through the Application of Visual Analytics." Currently, he is a visiting assistant professor at Purdue University working as a member of the visual analytics for command, control, and interoperability environments Department of Homeland Security Center of Excellence. His research interests include visual analytics, illustrative visualization, volume rendering, non-photorealistic rendering and geovisualization. Contact him at `rmacieje@purdue.edu`.

Printed in the United States
by Baker & Taylor Publisher Services